IE COMMONWEALTH AND INTERNATIONAL LIBRARY
int Chairmen of the Honorary Editorial Advisory Board
ROBERT ROBINSON, O.M., F.R.S., LONDON
AN ATHELSTAN SPILHAUS, MINNESOTA

LECTRICAL ENGINEERING
neral Editor: N. HIL

ALCULATIONS IN UNDAMENTAL PHYSICS

OLUME I: MECHANICS AND HEAT

W0110886

VOL. I: Mechanics and Heat

VOL. II: Electricity and Magnetism

VOL. III: Waves and Electronics

CALCULATIONS IN FUNDAMENTAL PHYSICS

In Three Volumes

VOLUME I: MECHANICS AND HEAT

by

T. HEDDLE, B.SC. (HONS.), A.INST.P.

Senior Lecturer, South East London Technical College

PERGAMON PRESS

Oxford · New York · Toronto

Sydney · Braunschweig

PERGAMON PRESS LTD.,
Headington Hill Hall, Oxford
PERGAMON PRESS INC.,
Maxwell House, Fairview Park, Elmsford, New York 10523
PERGAMON OF CANADA LTD.,
207 Queen's Quay West, Toronto 1
PERGAMON PRESS (AUST.) PTY. LTD.,
19a Boundary Street, Rushcutters Bay, N.S.W. 2011, Australia
VIEWEG & SOHN GMBH,
Burgplatz 1, Braunschweig

Copyright © 1971 Trevor Heddle

All Rights Reserved. No part of this publication may be reproduced, stored in a retrieval system, or transmitted, in any from or by any means, electronic, mechanical, photocopying, recording or otherwise, without the prior permission of Pergamon Press Ltd.

First edition 1971

Library of Congress Catalog Card No. 72-121012

Printed in Hungary

This book is sold subject to the condition that it shall not, by way of trade, be lent, resold, hired out, or otherwise disposed of without the publisher's consent, in any form of binding or cover other than that in which it is published.

08 015829 3

CITY OF LEICESTER POLYTECHNIC LIBRARY
53.0012
H
530.076
HED
14.7.81

CONTENTS

	Pages
INTRODUCTION	ix
REFERENCE DATA	xi

	Problem Nos.
1. *MOTION*	
Rectilinear Motion	1–9
Acceleration of Gravity	10–14
Parabolic Motion	15–20
Relative Velocity	21–25
Angular Motion	26–32
Non-uniform Acceleration	33–36
2. *COMBINATIONS OF FORCES*	
Polygons and Resolution	1–7
Centre of Gravity	8–14
Inclined Plane	15–18
Friction	19–23
Balanced Beams	24–29
Shearing Force and Bending Moment	30–33
3. *FORCE AND ACCELERATION*	
Horizontal or Vertical Motion	1–10
Momentum	11–15
Motion on Gradient	16–22
Coupled Objects	23–29
4. *ENERGY AND POWER*	
Work and Energy	1–8

Problem Nos.

Power	9–14
Movement on Gradient	15–20
5. *MACHINES*	
Lifting	1–7
Rotation and Power	8–10
Pulley Systems	11–16
Gears and Chains	17–21
6. *ELASTICITY*	
Helical Springs	1–4
Young's Modulus	5–10
Compound Bars	11–13
Bulk Modulus	14–16
Torsion	17–19
Stored Energy	20–24
7. *EXPANSION AND TEMPERATURE*	
Linear Temperature Scales	1–2
Linear and Superficial Expansion	3–11
Volumetric Expansion	12–20
Thermal Stress	21–24
8. *STATIC FLUIDS*	
Pressure in Fluids	1–6
Archimedes' Principle	7–14
Fortin Barometer	15–17
U-tube Manometers	18–24
Surface Tension	25–31
9. *IDEAL GASES*	
Constant Temperature	1–6
Constant Pressure or Volume	7–10

	Problem Nos.
Single Gas Equation	11–15
Change of Density	16–19
Ideal Gas Scale of Temperature	20–22
10. *QUANTITY OF HEAT*	
Specific Heat	1–8
Latent Heat	9–17
Heat and Work	18–23
Calorific Values	24–29
11. *CENTRIPETAL ACCELERATION*	
Centripetal Force and Acceleration	1–7
Gravitation and Circular Orbits	8–13
Banking and Vehicle Stability	14–19
12. *ROTARY INERTIA*	
Moment of Inertia	1–7
Rotational Energy	8–14
Change of Axis	15–20
13. *SIMPLE HARMONIC MOTION*	
Projected Rotation	1–7
Horizontal Elastic Motion	8–10
Simple Pendulum	11–16
Vertical Elastic Motion	17–20
Torsional and Compound Pendulums	21–26
14. *TRANSFER OF HEAT*	
Thermal Conduction	1–6
Newton's Law of Cooling	7–15
Cooling Correction	16–17
Radiation	18–24

Problem Nos.

15. *MOLECULAR THEORY OF GASES AND VAPOURS*
Avogadro's Law 1–7
Saturated Vapour 8–15
Relative Humidity 16–20
Pressure and Kinetic Energy 21–25
Temperature and Kinetic Energy 26–29

16. *FLUID FLOW*
Flow and Work 1–7
Bernoulli's Equation 8–15
Viscosity 16–22

17. *THERMODYNAMICS*
Work by Expanding Gas 1–6
Heat, Work, and Internal Energy 7–11
Principal Specific Heats of Gases 12–16
Adiabatic Changes 17–23
Reversible Cycles 24–31

Page

INDEX 291

INTRODUCTION

THIS volume is the first of three which form neither a complete textbook of physics nor a mere collection of exercises. It contains a sequence of worked examples and unworked exercises arranged to present a logical development from the basic definitions of physical quantities to the more complex relationships between them.

Each section starts with a worked example in which first a numerical problem is stated, then a general quantitative development of the relevant concepts and equations is usually given, and, finally, the numerical solution is shown in detail. The exercises which follow in each section refer to the particular topic of the worked example or examples which precede them, and do not range widely from this topic.

S.I. units are used almost exclusively, the exceptions being where other closely related units are more convenient. For example, in simple problems in statics use of a gravitational unit of force often avoids tedious arithmetic.

The data of most questions are stated to only two significant figures, but the data of the reference section are stated to greater accuracy in most cases. Most answers are therefore given to an accuracy of two significant figures, even when a third figure may appear to terminate the number exactly.

The divisions of the subject matter follow fairly traditional lines to help in finding a problem or explanation of a particular point. While such divisions make the portions readily digestible, this is not sufficient in a complete course in physics, which should obviously include considerable amounts of laboratory work, both demonstrations and experiments, and practice in the description and explanation of apparatus and other examp les of the embodiment of physical principles. The attempt has been made not to present topics as they might appear in specialist studies of specific branches of the subject, but to show the fundamental physical principles on which applied physics and engineering are founded.

The most elementary work of the book may be described as revision of the physics of pre-O.N.C. courses in science or engineering or of O-level G.C.E. work in physics, while the final level is somewhat above the physics of most O.N.C. and O.N.D. courses in science and engineering. Since syllabuses vary between different colleges, the range of topics is wider and some treatment is deeper than may be necessary for any particular course. The introductory notes should be useful for reference and for further work in H.N.C. and H.N.D. courses.

I gladly acknowledge help from many sources during the years in which this book has grown. Much assistance has been gained from the publications of numerous authors who may recognize ideas that they have expressed originally. Many colleagues have given helpful advice on specific points even unwittingly before the book was started. I regret any errors and obscurities which readers may find, and would welcome their being brought to my attention.

TREVOR HEDDLE

the fifth. Thus eqn. (1.1) may be used without involving s, or eqn. (1.2) without a. These are two of five equations connecting the five quantities four at a time, i.e. omitting one at a time. The other three may be obtained from eqns. (1.1) and (1.2) entirely algebraically, or with reference to a velocity–time graph.

Thus from eqn. (1.1), $t = (v-u)/a$, and from eqn. (1.2), $t = 2s/(u+v)$. Then eliminating t,

$$(v-u)/a = 2s/(u+v).$$

Cross-multiplying gives

$$v^2 - u^2 = 2as. \quad (1.3)$$

Alternatively, the velocity–time graph of Fig. 1.1 may be used. Here the velocity is shown rising with constant acceleration given by eqn. (1.1) and indicated on the graph by the gradient of the straight line. Equation (1.2) is also related to this graph, since $(u+v)/2$ is indicated by the average height of the graph, and t is indicated by the base length under the line. Therefore the area under the graph represents $t(u+v)/2$ which equals the distance s.

Calculating the same area in other ways gives other equations. Thus by adding the areas of the rectangle of height u and the

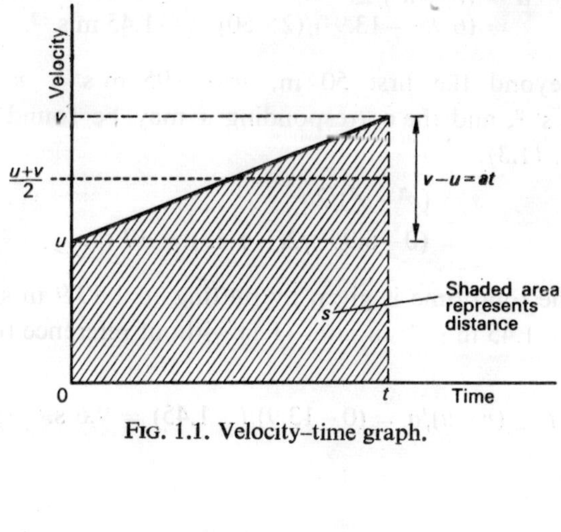

FIG. 1.1. Velocity–time graph.

triangle of height at,

$$s = ut + \tfrac{1}{2}at^2, \tag{1.4}$$

or subtracting the triangle from the rectangle of height v gives

$$s = vt - \tfrac{1}{2}at^2. \tag{1.5}$$

It can be seen that eqns. (1.1), (1.2), (1.3), (1.4), and (1.5) are distinguished by the omission of s, a, t, v, and u respectively, and the particular equation required for a calculation may be recognized by this.

Solution

First the data are converted to units of metres and seconds, as in ex. 1.1, and then the acceleration is calculated.

Thus for 50 km h^{-1},

$$u = 50/3.6 = 13.9 \text{ m s}^{-1}$$

and after a distance of $s = 50$ m,

$$v = 25/3.6 = 6.95 \text{ m s}^{-1}.$$

Then, from eqn. (1.3),

$$\begin{aligned} a &= (v^2 - u^2)/2s \\ &= (6.95^2 - 13.9^2)/(2 \times 50) = -1.45 \text{ m s}^{-2}. \end{aligned}$$

(a) Beyond the first 50 m, $u = 6.95$ m s^{-1}, $v = 0$, $a = -1.45$ m s^{-2}, and the corresponding s may be found without t from eqn. (1.3):

$$\begin{aligned} s &= (v^2 - u^2)/2a \\ &= (0 - 6.95^2)/2(-1.45) = 17 \text{ m}. \end{aligned} \tag{a}$$

(b) The total time is given by putting $u = 13.9$ m s^{-1}, $v = 0$, and $a = -1.45$ m s^{-2} in eqn. (1.1) without reference to distance. Thus

$$t = (v - u)/a = (0 - 13.9)/(-1.45) = 9.6 \text{ s}. \tag{b}$$

(c) During the 2 s interval, $a = -1.45$ m s^{-2} starting from $u = 6.95$ m s^{-1}. Then, without v, eqn. (1.4) gives

$$s = ut+\tfrac{1}{2}at^2 = (6.95\times 2)+\tfrac{1}{2}(-1.45)\times 2^2 = 11 \text{ m.} \qquad \text{(c)}$$

The answer (c) is confirmed by noting that the average velocity for the 2 s interval occurs after 1 s when $v = u-at = 6.95-1.45$ $=5.5$ m s^{-1}. Then, for $t = 2$, the distance $= 5.5\times 2 = 11$ m.

1.3. A car accelerates from 36 km h^{-1} to 72 km h^{-1} (a) in a time of 4.0 s, and (b) over a distance of 45 m. Calculate the acceleration in each case, the distance covered in case (a), and the time taken in case (b). [(a) 2.5 m s^{-2}, 60 m; (b) 3.3 m s^{-2}, 3.0 s.]

1.4. A vehicle accelerates at the rate of 1.0 m s^{-2}. Calculate the velocity in km h^{-1} reached at a distance of 30 m beyond the point at which the velocity is 7.0 m s^{-1}. [38 km h^{-1}.]

1.5. A ball rolls a certain distance at a speed of 5.0 m s^{-1} and is then kicked forward so as to travel a further equal distance in the same direction at a speed of 15.0 m s^{-1}. Calculate the average velocity. [7.5 m s^{-1}.]

1.6. A vehicle has constant acceleration for the first half of a certain distance and constant velocity for the second half. Calculate the average velocity if the initial and final velocities are 20 km h^{-1} and 60 km h^{-1}. [48 km h^{-1}.]

1.7. A car travels 1.0 km from rest to rest with constant acceleration for the first 0.25 km and constant deceleration for the remaining 0.75 km. Calculate the time taken (a) if the maximum velocity reached is 50 km h^{-1}, and (b) if the acceleration and deceleration are double those required for (a). [(a) 2.4 min; (b) 1.7 min.]

1.8. A vehicle takes 35 m to stop in an emergency from an initial speed of 63 km h^{-1}. One-fifth of the time elapses with the vehicle running at constant velocity before the driver reacts to the situation. Assuming constant deceleration when the brakes are applied, calculate (a) the reaction time, and (b) the distance covered during this time. [(a) 0.67 s; (b) 12 m.]

1.9. A car accelerates from rest to a speed of 40 km h^{-1} in a distance of 50 m and then continues at constant speed. Using a velocity–time graph, or otherwise, calculate how long it takes to catch up with the car on a motor-cycle starting from rest 2.0 min later than the car and reaching twice the speed with twice the acceleration. [2.1 min.]

Acceleration of Gravity

1.10. Worked Example

A ball thrown vertically upwards leaves the hand at a height of 2.0 m and reaches a maximum height of 15 m from the ground. Calculate (a) the initial velocity, and (b) the time taken to return to a height of 2.0 m.

Introduction

When an object moves in any direction under the action of the force of gravity alone, it has a downward acceleration which is independent of its mass. In practice various additional forces may alter the acceleration; e.g. air resistance reduces the acceleration with which an object falls to an extent depending on the shape and velocity of the object. Also any considerable change in the distance from the centre of the earth affects the acceleration of gravity. However, in many cases of objects thrown or falling in air near the earth's surface, it may be assumed that gravity produces a constant downward acceleration g relative to the earth's surface; $g = 9.81$ m s^{-2}.

Solution

(a) Taking the upward direction as positive, the maximum height $s = 13$ m above the hand, when the velocity $v = 0$ instantaneously. Still taking the upward direction as positive, the acceleration of gravity must be negative, i.e. $a = -9.81$ m s^{-2}.

REFERENCE DATA

Substance	Density at s.t.p. (kg m^{-3})	Temp. coef. of exp. (0–150 °C) (deg C^{-1})	Specific heat (0–100°C) (J kg^{-1} deg C^{-1})	Young's modulus (N m^{-2})	Thermal conductivity (W m^{-1} deg C^{-1})
Air	1.29				0.023
Aluminium	2.7×10^{3}	$\alpha = 26\times10^{-6}$	910	7.0×10^{10}	200
Asbestos					0.15
Brass	8.5×10^{3}	$\alpha = 19\times10^{-6}$	370	9.0×10^{10}	
Carbon dioxide	1.98				
Copper	8.9×10^{3}		390		380
Glass		$\alpha = 8.0\times10^{-6}$			0.80
Hydrogen	0.090				
Invar		$\alpha = 1\times10^{-6}$			
Iron	7.85×10^{3}	$\alpha = 12\times10^{-6}$	470		
Mercury	13.6×10^{3}	$\gamma = 182\times10^{-6}$			
Oxygen	1.43				
Paraffin oil		$\gamma = 9.0\times10^{-4}$	2200		
Steel	7.7×10^{3}	$\alpha = 12\times10^{-6}$	460	2.2×10^{11}	
Water	1000		4190		

Latent heat of ice at 0°C $= 3.35\times10^{5}$ J kg^{-1}

Latent heat of steam at 100°C $= 2.25\times10^{6}$ J kg^{-1}

Pressure of saturated water vapour (mmHg)	9.2	17.5	31.8	55.3	149	289	526
Temperature (°C)	10	20	30	40	60	75	90

General Constants

Standard acceleration of gravity $g = 9.81\ \text{m s}^{-2}$
Avogadro's number $N_A = 6.02\times10^{26}$ molecules per kilomole
Universal gas constant $R = 8.31\times10^{3}\ \text{J K}^{-1}$ per kilomole
Ideal gas volume at s.t.p. = $22.4\ \text{m}^3$ per kilomole

s.t.p. = standard temperature and pressure
= 0°C and 760 mmHg
= 273.15 K and $1.013\times10^{5}\ \text{N m}^{-2}$

Stefan–Boltzmann constant $\sigma = 5.67\times10^{-8}\ \text{W m}^{-2}\ \text{K}^{-4}$
Wien's displacement constant = 2.9×10^{-3} m K
Velocity of light in vacuum $c = 3.00\times10^{8}\ \text{m s}^{-1}$

1 litre = $10^{-3}\ \text{m}^3$
1 tonne = 10^3 kg

CHAPTER 1

MOTION

Rectilinear Motion

1.1 Worked Example

The velocity of a vehicle is reduced steadily from 54 km h^{-1} to 18 km h^{-1} in a time of 5.0 s. Calculate (a) the deceleration, and (b) the distance covered.

Introduction

Deceleration is the opposite of acceleration: deceleration is the rate of decrease of velocity with time, whereas acceleration is the rate of increase. Both are given by a in the following equation, where a is positive for acceleration and negative for deceleration. Thus when the velocity changes from u to v at a constant rate and in a time t,

$$a = (v-u)/t. \tag{1.1}$$

The average velocity is the total distance s divided by the total time t taken to cover that distance. In the particular case of constant acceleration, positive or negative, the average velocity is also the mean of the initial and final velocities. So, for constant acceleration,

$$s/t = (u+v)/2. \tag{1.2}$$

Note that the positive direction must be the same for each quantity u, v, s, and a, and that s as an algebraic symbol in an equation refers to distance but s as a unit refers to a second.

Different units of distance or of time must not be used together in these equations. Thus when metres (m) and seconds (s) are used, velocities in kilometres per hour (km h^{-1}) must be converted to metres per second (m s^{-1}):

$$1 \text{ km h}^{-1} = (1000 \text{ m})/(3600 \text{ s}) = (1/3.6) \text{ m s}^{-1}.$$

The corresponding unit of acceleration is then seen from eqn. (1.1) to be (metre per second) per second, abbreviated to m s^{-2}.

Solution

Using metres and seconds,

$$u = 54/3.6 = 15 \text{ m s}^{-1}$$

and

$$v = 18/3.6 = 5 \text{ m s}^{-1}.$$

Then eqn. (1.1) gives

$$a = (v-u)/t = (5-15)/5.0 = -2.0 \text{ m s}^{-2},$$

i.e. a negative acceleration, or a deceleration of 2.0 m s^{-2}. (a)

Lastly, eqn. (1.2) gives

$$s/5.0 = (u+v)/2 = (15+5)/2.$$

$$\therefore s = 50 \text{ m}. \quad \text{(b)}$$

1.2. Worked Example

A vehicle under constant deceleration travels a distance of 50 m while the velocity drops from 50 km h^{-1} to 25 km h^{-1}. Calculate (a) the further distance, and (b) the total time taken in coming to rest. Also calculate (c) the distance covered during an interval of 2.0 s directly after the speed has dropped to 25 km h^{-1}.

Introduction

In many problems concerning the quantities u, v, a, t, and s, as defined in ex. 1.1, an equation is required for finding one unknown quantity using three known quantities, without involving

Then, using eqn. (1.3),

$$u^2 = v^2 - 2as = 0 - 2(-9.81)13 = 255$$

$$\therefore\ u = 16 \text{ m s}^{-1} \quad \text{(2 sig. fig.).} \qquad \text{(a)}$$

(b) Since the velocity changes by 9.81 m s^{-1} in every second, it takes a time of 16/9.81 s for the initial velocity of 16 m s^{-1} to be reduced to zero during upward movement and a further equal time for the ball to fall back with increasing downward velocity exactly reversing the previous upward movement. Therefore the time required to return to the initial height

$$= 2 \times 16/9.81 = 3.3 \text{ s} \quad \text{(2 sig. fig.).} \qquad \text{(b)}$$

Alternatively, for an algebraic solution for (b), eqn. (1.4) may be used. This equation will apply to both the upward and the downward stages of the motion, if s is the height above the initial level, even after the maximum height has been reached.

Thus $s = ut + \frac{1}{2}at^2 = 16t + \frac{1}{2}(-9.81)\,t^2 = t(16 - 4.9t)$. Therefore when $s = 0$, $t = 0$ or $t = 16/4.9 = 3.3$ s, i.e. the ball is at a height, $s = 0$ both initially and 3.3 s later.

1.11. With what speed must a stone be thrown vertically upwards for it to remain in the air for a time of 3.0 s? What is the maximum height reached? [14.7 m s^{-1}; 11 m.]

1.12. In a shock test it is required to simulate the effect of a head-on collision between a light object and a massive one, each travelling at 50 km h^{-1}. Calculate the height from which the light object should be dropped on to the other to obtain the required impact. [39 m.]

1.13. What maximum height is reached by a ball thrown upwards if its height is 10.0 m after a time of 2.00 s? [11.2 m.]

1.14. A ball is thrown vertically upwards at a speed of 9.8 m s^{-1}. One second later another ball is thrown likewise. Calculate the height of collision. [3.7 m.]

Parabolic Motion

1.15. Worked Example

A jet of water from a hose fixed at ground level rises at an angle of 20° to the horizontal and has a range of 25 m. Calculate (a) the speed with which the water leaves the hose, and (b) the maximum height reached.

Introduction

The motion, with initial velocity u, as shown in Fig. 1.2, is resolved into two components, namely a horizontal component which continues with a constant velocity of $u \cos \theta$ over the full

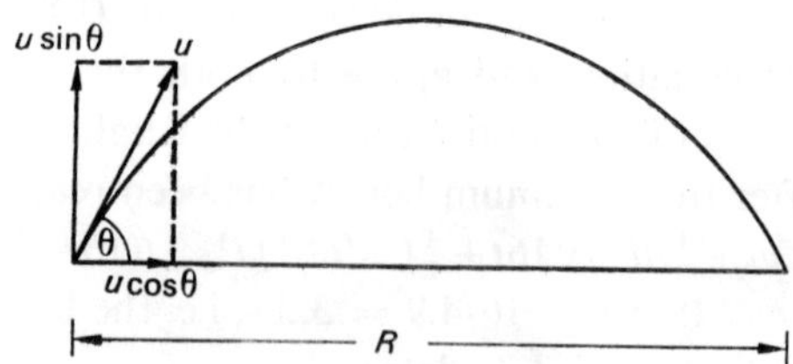

FIG. 1.2. Initial velocity components and trajectory.

range, assuming negligible air resistance, and a vertical component which commences with a velocity of $u \sin \theta$, and is governed by the acceleration due to gravity, still assuming negligible air resistance.

Horizontally, the range

$$R = u \cos \theta \; t, \tag{1.6}$$

where t = time of complete motion. The vertical velocity $u \sin \theta$ is reduced to zero at half time, $t/2$, and is completely reversed at the end of time t. The vertical acceleration $a = -g$, where g is the acceleration due to gravity and the upward direction is taken to be positive, as for $u \sin \theta$. Then, from eqn. (1.1), $-g = (0-u \sin \theta)/(t/2)$.

$$\therefore\ gt/2 = u \sin \theta. \tag{1.7}$$

Solution

(a) From eqn. (1.6), with $\cos 20° = 0.94$,

$$25 = u0.94t \tag{1.8}$$

and from eqn. (1.7), with $\sin 20° = 0.342$ and $g = 9.81 \text{ m s}^{-2}$, $9.81t/2 = u0.342$. Then substituting for t in eqn. (1.8),

$$25 = (u0.94)(u0.342\times 2/9.81).$$

$$\therefore\ u^2 = (25\times 9.81)/(0.94\times 0.342\times 2) = 381.$$

$$\therefore\ u = 19.5 = 20 \text{ m s}^{-1} \quad \text{(2 sig. fig.)}. \qquad \text{(a)}$$

(b) In the vertical motion, the maximum height s is reached when $v = 0$ and $u \sin \theta$ replaces u in eqn. (1.3);

$$v^2 = (u \sin \theta)^2 + 2as,$$

i.e.

$$0 = (19.5\times 0.342)^2 + 2(-9.81)s.$$

$$\therefore\ s = (19.5\times 0.342)^2/19.62 = 2.3 \text{ m}. \qquad \text{(b)}$$

1.16. A ball is kicked from the ground and rises at 30° to the horizontal with velocity 15 m s^{-1}. Calculate the peak height of the trajectory, the time of flight, and the horizontal range. [2.9 m; 1.5 s; 20 m.]

1.17. Two horizontal benches of different heights have a gap between them. A small box slides freely off the higher bench and lands on the lower. Calculate the minimum velocity which the box must have as it reaches the gap (a) when the gap width is 1.0 m and the difference of height is 0.20 m, (b) when the width is the same as that of (a) but the difference of height is doubled, and (c) when both dimensions are double those of (a). [(a) 5.0 m s^{-1}; (b) 3.5 m s^{-1}; (c) 7.0 m s^{-1}.]

1.18. Calculate the maximum horizontal range and the corresponding time of flight and height of trajectory for a ball leaving the ground at a speed of 20 m s^{-1} with negligible air resistance. [41 m; 2.9 s; 10 m.]

1.19. From the top of a vertical cliff, 10 m high, a stone is thrown (a) horizontally, and (b) upwards at 45° to the horizontal. Calculate the initial velocity in each case for the stone to reach the sea at a point 10 m from the base. [(a) = (b); 7 m s^{-1}.]

1.20. A ball is thrown upwards at a speed of 25 m s^{-1} at an angle of 50° to the horizontal from a height of 2.0 m above ground. Calculate its position, direction of movement, and speed after a time of 3.0 s, assuming negligible air resistance. [15 m above ground, 48 m horizontally; 32° to the horizontal; 19 m s^{-1}.]

Relative Velocity

1.21. WORKED EXAMPLE

A boat is rowed across a river 0.50 km wide. The velocity of the boat relative to the water is 6.0 km h^{-1} and the water flows uniformly downstream at a rate of 3.0 km h^{-1}. Determine (a) the time of crossing, the position of landing, and the resultant velocity when the boat heads directly across the river, and (b) the time of crossing and the direction in which the boat must head to reach a point directly opposite the point of departure.

Solution

(a) When the boat heads directly across, the flow has no effect on the forward velocity; so the boat crosses at 6.0 km h^{-1} for 0.50 km through the water. Then the time of crossing = 0.50/6.0 $= \frac{1}{12}$ h.

Throughout this time the boat is carried downstream at 3.0 km h^{-1}, and therefore lands at a distance downstream of $3.0 \times \frac{1}{12}$ $= \frac{1}{4}$ km.

Figure 1.3 shows the distance travelled:

$$AB = \sqrt{(0.5^2+0.25^2)} = 0.56 \text{ km}.$$

Therefore the resultant velocity $= 0.56/(\frac{1}{12}) = 6.7 \text{ km h}^{-1}$.

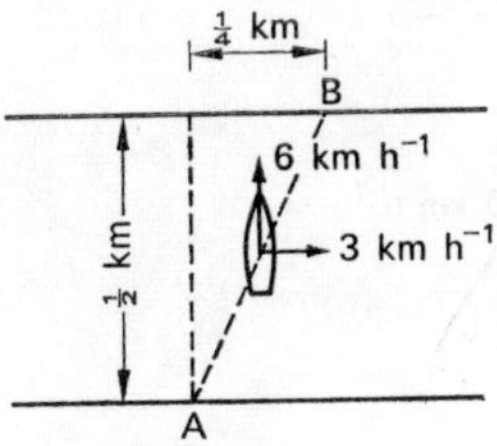

FIG. 1.3. Ex. 1.21(a).

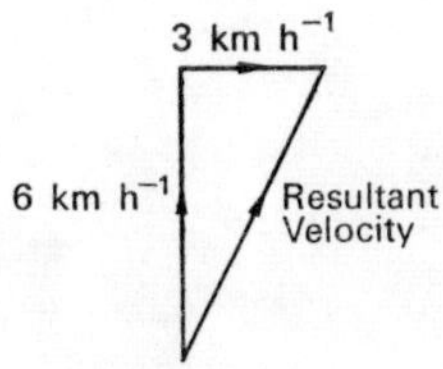

FIG. 1.4. Velocity diagram for ex. 1.21(a).

Alternatively, the resultant velocity may be found by adding velocities by means of a triangle (or parallelogram) of velocities in which the length and direction of each side represents the magnitude and direction of one of the velocities. Thus Fig. 1.4 shows the velocity of the boat relative to the water added to that of the water relative to the bank to give the resultant velocity of the boat relative to the bank. Note that the component velocities follow one another around two sides of the triangle, and the resultant "takes the short cut" along the third side.

(b) For the resultant velocity to be directed perpendicular to the bank, the boat must head obliquely upstream so that the upstream component of its velocity just offsets the flow of 3.0 km h^{-1} Thus in Fig. 1.5, $6 \cos \theta = 3$, and the resultant velocity $= 6 \sin \theta$. Then $\cos \theta = \frac{1}{2}$, and $\theta = 60°$. So the boat must head upstream at 60° to the bank.

The time taken to cross the 0.5 km from A to C at 6 sin 60° km h^{-1} $= 0.5/(6 \sin 60°) = 0.5/(6\times0.866) = 0.096 \text{ h} \simeq 6 \text{ min}$.

1.22. A train 100 m long, travelling at 10 m s^{-1}, passes a second train, 80 m long, travelling at 20 m s^{-1} on a parallel track. Calcu-

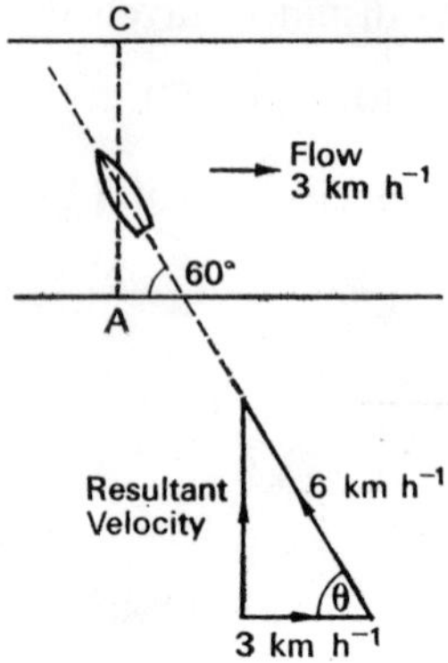

FIG. 1.5. Ex. 1.21(b).

late the time during which their lengths overlap when travelling (a) in opposite directions, and (b) in the same direction. [(a) 6.0 s; (b) 18 s.]

1.23. A ship heading to the north-east moves at 10 km h^{-1} relative to the sea, which flows at 4.0 km h^{-1} to the west relative to the shore. Determine (a) the direction and velocity at which the ship moves relative to the shore, and (b) the direction in which the ship must head with a speed of 10 km h^{-1} to move north-east relative to the shore. [(a) 7.7 km h^{-1}, N 23° E; (b) N 61° E.]

1.24. A cyclist riding at 20 km h^{-1} due west finds the wind to be from the north-west. When he rides at 20 km h^{-1} due east, he finds the wind to be from the north-north-east. Determine the true direction of the wind relative to the ground. [N 16° W.]

1.25. A ship *A*, which moves due west at 16 km h^{-1}, is 5.0 km due south of a ship *B*, which moves due south at 12 km h^{-1}. Determine (a) the velocity of *B* relative to *A*, and (b) the time taken to reach their shortest distance apart. [(a) 20 km h^{-1}, E 37° S; (b) 9 min.]

Angular Motion

1.26. Worked Example

The velocity of a rolling wheel of diameter 30 cm increases steadily from 24 m s^{-1} to 36 m s^{-1} in a time of 2.2 s. Calculate (a) the total number of revolutions turned in this time, and (b) the angular acceleration.

Introduction

Angular acceleration is normally stated in units of radians second^{-2} (rad s^{-2}), the radian being defined with reference to Fig. 1.6, where s = length of circular arc, r = radius, θ = angle in radians subtended at the centre. $\theta = 1$ rad when $s = r$, and in general $\theta = s/r$ rad.

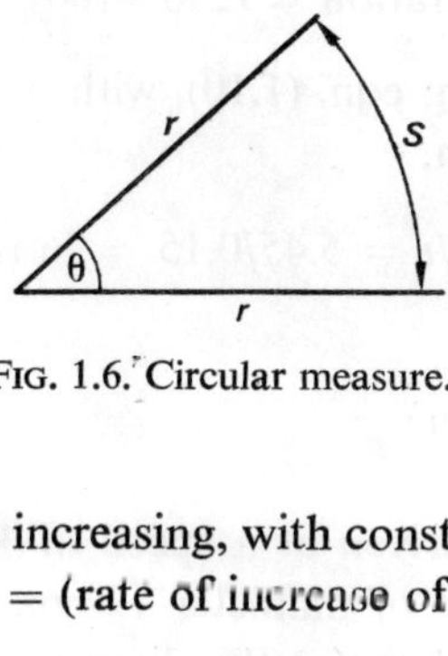

FIG. 1.6. Circular measure.

Now if θ and s are increasing, with constant r and fixed centre, (rate of increase of θ) = (rate of increase of s)/r, i.e.

$$\omega = v/r, \tag{1.9}$$

where ω = angular velocity in rad s^{-1}, v = linear velocity at constant radius r from the fixed centre, and the unit of length must be the same for both r and v. Further, from eqn. (1.9), if ω increases, (rate of increase of ω) = (rate of increase of v)/r, i.e.

$$\alpha = a/r, \tag{1.10}$$

where α = angular acceleration in rad s^{-2}, a = linear acceleration at constant radius from the fixed centre, and again the units must be consistent for a and r.

Solution

(a) The average linear velocity = $(24+36)/2 = 30$ m s^{-1}.

$\therefore$ the distance rolled in 2.2 s = $2.2\times30 = 66$ m.

Now when the wheel rolls one revolution, the length of its track along the ground equals the circumference. Therefore for n revolutions, $n(0.3\pi)$ m = 66 m.

$$\therefore n = 66/0.3\pi = 70. \quad \text{(a)}$$

(b) The angular velocity changes from $24/0.15 = 160$ rad s^{-1} to $36/0.15 = 240$ rad s^{-1} in a time of 2.2 s.

$$\therefore \text{angular acceleration} = (240-160)/2.2 = 36 \text{ rad s}^{-2}. \quad \text{(b)}$$

Alternatively, using eqn. (1.10), with $a = (36-24)/2.2 = 5.45$ m s^{-2}, and $r = 0.15$ m,

$$\alpha = a/r = 5.45/0.15 = 36 \text{ rad/s}^{-2}. \quad \text{(b)}$$

1.27. Worked Example

A shaft is to be driven at a speed of 150 rev min^{-1} by belt drive from a pulley A of diameter 12 cm, turning at a speed of 200 rev min^{-1}. Calculate (a) the diameter of the pulley B to be fitted to the shaft. Then if the pulley A accelerates uniformly from 200 rev min^{-1}, so as to rotate 300 rev in 30 s, calculate (b) the angular acceleration of the shaft.

Introduction

The five equations of motion for constant acceleration, given in exs. 1.1 and 1.2 apply both to linear quantities and to the corresponding angular quantities. Thus eqn. (1.4) is useful here:

$$s = ut+\tfrac{1}{2}at^2, \qquad \text{for linear motion,}$$

$$\theta = \omega_1 t+\tfrac{1}{2}\alpha t^2, \qquad \text{for angular motion,}$$

where θ is the angle turned in a time t with angular acceleration α from an initial angular velocity ω_1.

Also note that if the belt does not slip (as may be assumed) the circumferences of both pulleys must have the same linear movement as the belt, i.e. the same linear displacement, velocity, and acceleration.

Solution

(a) As in ex. 1.26, $\omega = v/r$. Then, equating the linear velocities of the two pulleys, $v = r_A\omega_A = r_B\omega_B$.

For each wheel $r = d/2$, where d = diameter, and $\omega = n2\pi/60$, where n = speed in rev s^{-1}. $\therefore\ d_A n_A = d_B n_B$

$$\therefore\ d_B = d_A n_A/n_B = 12\times 200/150 = 16 \text{ cm.} \qquad \text{(a)}$$

(b) Now for pulley A, θ = angle turned = $300\times 2\pi$ rad in time $t = 30$ s, and ω_1 = initial angular velocity = $200\times 2\pi/60$ = $20\pi/3$ rad s^{-1}.

Then, using $\theta = \omega_1 t+\frac{1}{2}\alpha t^2$,

$$600\pi = (20\pi/3)\times 30+\tfrac{1}{2}\alpha 30^2.$$

$\therefore$ angular acceleration of A, $\alpha_A = (600\pi-200\pi)/450 = 8\pi/9$.

Since both pulleys have the same linear acceleration, $a = r_A\alpha_A = r_B\alpha_B$,

$$\therefore\ \alpha_B = (r_A/r_B)\,\alpha_A = (\tfrac{12}{16})(8\pi/9)$$

$$= 2\pi/3 \text{ rad s}^{-2}. \qquad \text{(b)}$$

1.28. A rotor of diameter 20 cm accelerates from rest to a speed of 300 rev min^{-1} in a time of 10 s. Calculate (a) the number of revolutions made, (b) the angular acceleration, and (c) the final linear velocity of the rim. [(a) 25; (b) π rad s^{-2}; (c) π m s^{-1}.]

1.29. Two pulley wheels, 15 cm and 10 cm in diameter respectively, are coupled by a belt. The larger wheel is accelerated uniformly from 120 rev min^{-1} to 180 rev min^{-1} in a time of 20 s. Calculate the angular acceleration of each wheel and the angle turned during the 20 s by each wheel. [0.31 rad s^{-2}; 0.47 rad s^{-2}; 310 rad; 470 rad.]

1.30. The armature of a motor slows down from a speed of 3000 rev min^{-1} to rest in a time of 30 s after switching off. Calculate (a) the number of revolutions made in coming to rest, and (b) the angle turned during the third second. [(a) 750 rev; (b) 290 rad.]

1.31. A wheel of diameter 50.0 cm rolls without slipping at a rate of 600 rev min^{-1}. Calculate the linear velocity relative to the ground, at the following points on the rim of the wheel (a) at the top, (b) at 90° from the top, and (c) at 135° from the top. [(a) 31.4 m s^{-1}; (b) 22.2 m s^{-1}; (c) 12 m s^{-1}.]

1.32. A vehicle with wheels of diameter 0.50 m is brought to rest steadily in 50 m from a speed of 10 m s^{-1}. Calculate (a) the initial angular velocity of the wheels, (b) their angular deceleration, (c) the initial linear velocity, relative to the ground, of a point on the circumference of a wheel and at the same height as the centre, and (d) the angle turned by each wheel during the tenth second. [(a) 40 rad s^{-1}; (b) 4.0 rad s^{-2}; (c) 14 m s^{-1}; (d) 2.0 rad.]

Non-uniform Acceleration

1.33. Worked Example

Figure 1.7 shows the path *OABCDE* of a ball rolling into a hollow on damp ground, and such that the ball comes to rest at *E* after rising to *C* and rolling back. The following figures give the distances from *O* at regular intervals of time.

Time (s)	0	0.1	0.2	0.3	0.4	0.5	0.6	0.7	0.8	0.9	1.0
Distance (cm)	0	5	10.1	15.8	22.6	30.0	35.2	35.3	34.2	33.8	33.8

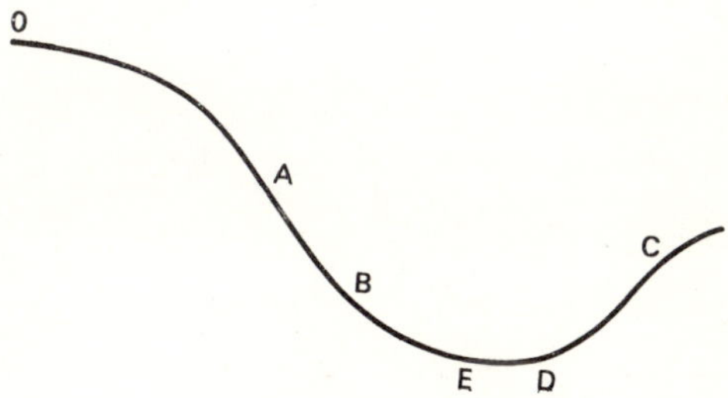

FIG. 1.7. EX. 1.33.

Obtain an approximate velocity–time graph and check that the area under this graph represents the distance covered. From the graph determine approximately the maximum values of (a) forward velocity, (b) reverse velocity, and (c) deceleration.

Introduction

When an object moves with varying acceleration, the eqns. (1.1), (1.2), (1.3), and (1.4) cannot be used since there is no constant value of a and $(u+v)/2$ is not necessarily equal to the average speed. However, such movement can be treated graphically, as in Fig. 1.8, where the data of this example are plotted.

The velocity at any instant is shown by the gradient of this distance–time graph. Thus where the graph is horizontal for a considerable time, as at E, the distance remains constant for that time, showing the velocity to be zero. Similarly, at C the velocity just reaches zero instantaneously.

At B, where the graph is steepest, the distance increases most rapidly and so the velocity is a maximum.

The velocities required to plot the velocity–time graph of Fig. 1.9 are found from the gradients at corresponding points of Fig. 1.8. Thus in Fig. 1.9 the velocity at *A* is shown to be 62 cm s^{-1}, which

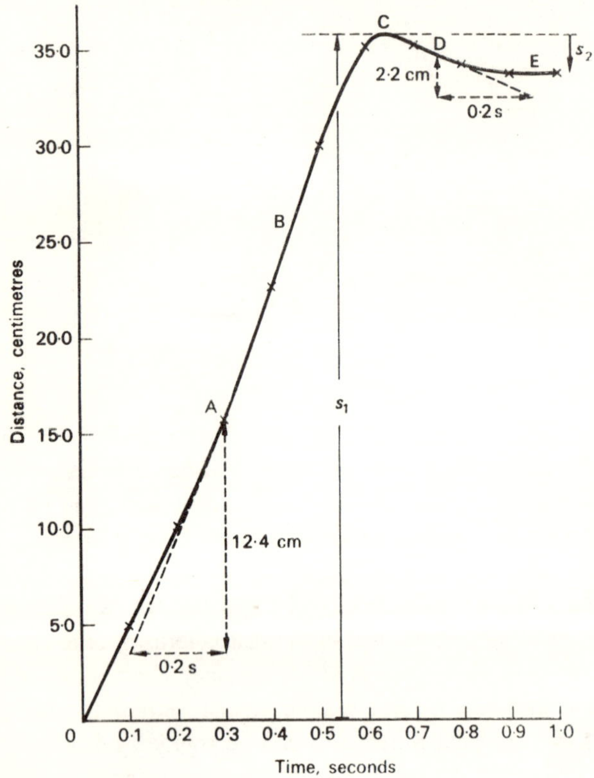

FIG. 1.8. **Distance–time graph for ex. 1.33.**

equals the gradient (12.4 cm)/(0.2 s) shown in Fig. 1.8. Similarly, at *D* the velocity is -11 cm s^{-1}, which is negative since the gradient of the graph of Fig. 1.8 is downward at *D*.

Each narrow strip under the velocity–time graph has an average height equal to the average velocity during the interval shown by the width of the strip. Therefore the area of each strip represents the distance covered in that interval, and as in Fig. 1.1, the total area under the line of the velocity–time graph represents the total distance covered. The positive forward distance s_1, and the negative backward distance s_2 are shown in both Figs. 1.8 and 1.9.

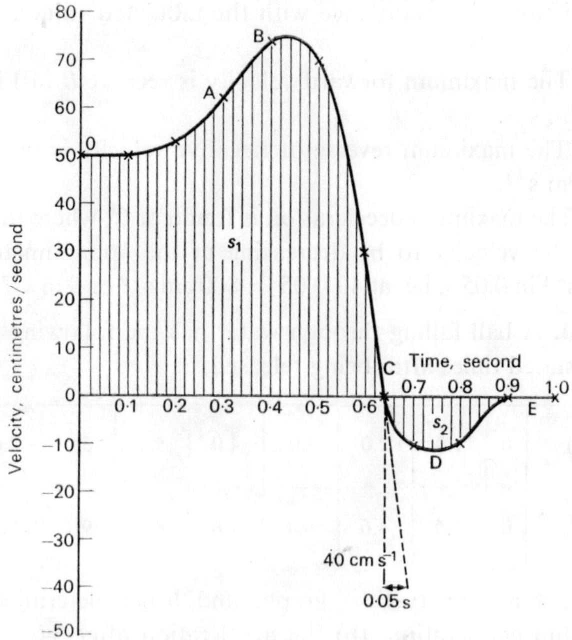

FIG. 1.9. Velocity–time graph for ex. 1.33.

The acceleration at any instant is shown by the gradient of the velocity–time graph. Thus maximum acceleration occurs where the velocity–time graph rises most steeply, and maximum deceleration where it drops most rapidly.

Solution

By counting squares on graph paper, or using the mid-ordinate rule, the area s_1 of Fig. 1.9 is found to be equivalent to $35\frac{1}{2}$ squares each representing a distance (10 cm s^{-1})×(0.1 s).

$$\therefore\ s_1 = 35\tfrac{1}{2}\times(10\times0.1) = 35\tfrac{1}{2}\text{ cm}.$$

Similarly, s_2 is equivalent to 2 squares.

$$\therefore\ s_2 = 2\times(10\times0.1) = 2\text{ cm},$$

approximately in accordance with the tabulated values, and with Fig. 1.8.

(a) The maximum forward velocity is seen at B in Fig. 1.9 to be 75 cm s^{-1}.

(b) The maximum reverse or negative velocity is seen at D to be 11 cm s^{-1}.

(c) The maximum deceleration is found at C, where the tangent shows the velocity to be decreasing at the approximate rate of 40 cm s^{-1} in 0.05 s, i.e. at 40/0.05 = 800 cm s^{-2} or 8 m s^{-2}.

1.34. A ball falling through water has the following velocities at the stated times after being released:

Time (s)	0	1.0	2.0	3.0	4.0	5.0	6.0	7.0	8.0
Velocity (m s^{-1})	0	2.5	4.6	6.4	7.6	8.5	9.1	9.4	9.6

Construct a velocity–time graph, and hence determine (a) the maximum acceleration, (b) the acceleration after 3.0 s from the start, and (c) the total distance fallen in the 8 s. [(a) 2.5 m s^{-2}; (b) 1.5 m s^{-2}; (c) 53 m.]

1.35. The following figures show the distances covered by a vehicle at one-minute intervals:

Time (min)	0	1.0	2.0	3.0	4.0	5.0	6.0	7.0	8.0	9.0	10.0	11.0	12.0
Distance (km)	0	0.5	1.0	1.5	1.9	2.0	2.0	2.0	2.2	2.9	3.6	3.8	3.8

Draw a distance–time graph, and hence obtain approximately the corresponding velocity–time graph. Check that the two areas under the latter graph represent the distances covered in the two stages of the journey. Determine (a) the average velocity during the 12 min, and (b) the approximate maximum velocity. [(a) 19 km h^{-1}; (b) 50 km h^{-1}.]

1.36. A pendulum bob swings so that its distance in centimetres from its lowest point is given by $3.5 \cos 2\pi t$, where t is the time in seconds measured from one extreme position. Determine (a) the maximum velocity, (b) the maximum acceleration, (c) the velocity when the distance is half its maximum value, and (d) the acceleration when the velocity is half its maximum value. [(a) 22 cm s^{-1}; (b) 140 cm s^{-2}; (c) 19 cm s^{-1}; (d) 120 cm s^{-2}.]

CHAPTER 2

COMBINATIONS OF FORCES

Polygons and Resolution

2.1. Worked Example

Figure 2.1 shows five forces acting in the same plane with magnitudes given in units of newtons (N). Determine the magnitude and direction of the resultant force.

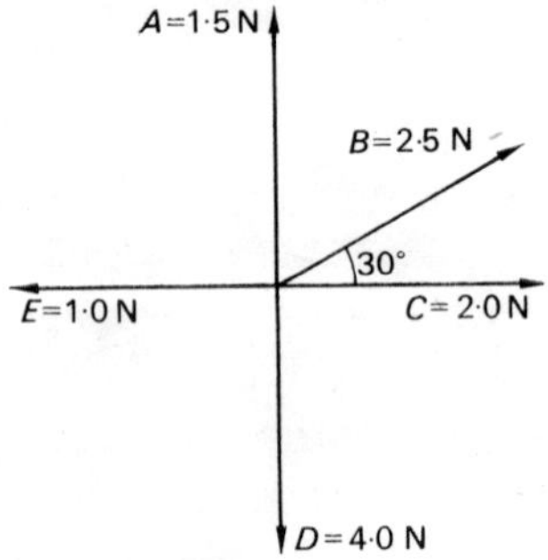

FIG. 2.1. Coplanar forces acting at a point (ex. 2.1).

Introduction

It is explained in exs. 3.1 and 3.2 that the fundamental unit of force in the M.K.S. system is the newton (N). In the present problem the magnitude of this unit is of no consequence since no comparison with other units is involved, but it may be useful to imagine the magnitude of one newton as roughly equal to the force required to support an object of mass about $\frac{1}{4}$ lb or about $\frac{1}{10}$ kg.

When two or more forces act on an object, with their directions in one plane and crossing at one point, they are equivalent to one resultant force. This resultant may be found by drawing a polygon, of which the sides are parallel to the directions of the forces and have lengths proportional to their magnitudes (as in Fig. 2.3). If the component forces *B*, *C*, *D*, etc., are shown following one another around the polygon, like parts of a continuous journey, then the resultant is shown by the length and direction of the equivalent journey "as the crow flies". If the component forces form a closed polygon without a resultant, then the resultant is zero and the component forces must be in equilibrium.

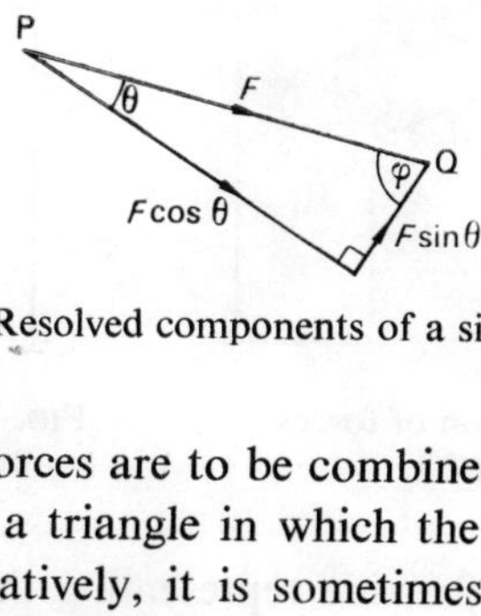

FIG. 2.2. Resolved components of a single force.

When only two forces are to be combined (as in Fig. 2.8), the polygon of forces is a triangle in which the third side represents the resultant. Alternatively, it is sometimes more convenient to draw a parallelogram of forces with a diagonal dividing it into two equivalent triangles. Then the diagonal represents the resultant in either triangle, and the components may be shown running through a common point with the resultant (as in Fig. 2.9).

A right-angled triangle is often used for the opposite purpose, namely that of resolving a given force (resultant) into two perpendicular component forces which are equivalent to the given force. Thus Fig. 2.2 shows two components $F \cos \theta$ and $F \sin \theta$, one following the other from *P* to *Q* and equivalent to the single force *F* shown by the straight line from *P* to *Q*. $F \sin \theta$ may be calculated alternatively as $F \cos \varphi$, where $\theta+\varphi = 90°$ in a right-angled triangle. Thus each perpendicular component equals the product of F and the cosine of the angle between that component and F.

Solution

Four solutions (a), (b), (c), and (d), are given below.

(a) The five forces may be taken in any order and represented by five sides of a six-sided polygon, as in Fig. 2.3 which shows the forces in clockwise sequence starting with B. Measurement

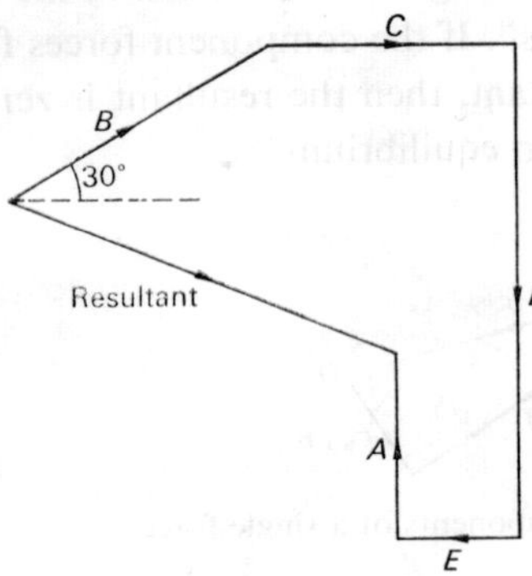

FIG. 2.3. Polygon of forces for ex. 2.1.

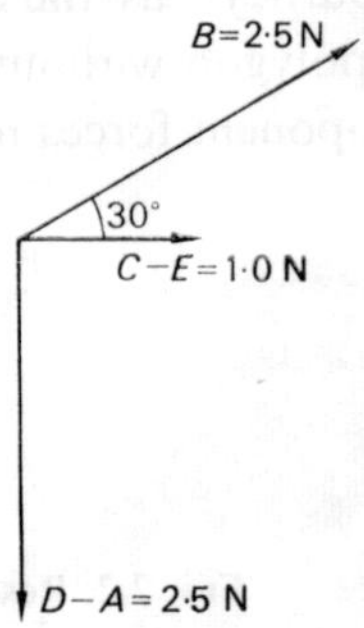

FIG. 2.4. Reduction of a number of forces (ex. 2.1).

shows the length of the side representing the resultant to be 3.4 times that representing E. The resultant is therefore 3.4 N at the inclination shown, i.e. approximately $21\frac{1}{2}°$ from the direction of C.

(b) The number of forces may first be reduced by combining those directly opposite each other. Thus in Fig. 2.4 E and C are shown by their resultant of 1.0 N, and A and D by their resultant of 2.5 N. Then these are shown with B in the polygon of Fig. 2.5, which gives the same resultant as in (a), Fig. 2.3.

(c) Figure 2.6 shows another six-sided polygon equivalent to Fig. 2.3 but taking the five original forces in a different order, namely C, B, A, E, D.

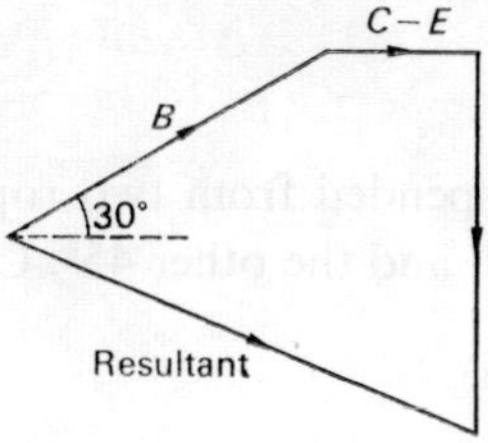

FIG. 2.5. Reduction of polygon of forces (ex. 2.1).

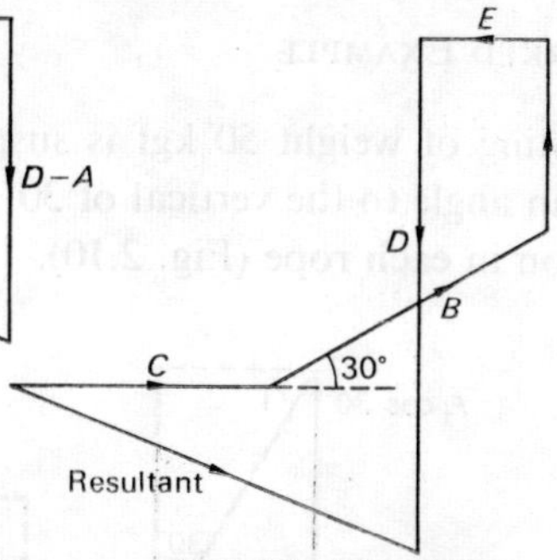

FIG. 2.6. Alternative version of Fig. 2.3.

(d) Lastly, the resultant may be calculated without drawing to scale. Thus B may be resolved into components parallel to A and C, as in Fig. 2.7 which is similar in principle to Fig. 2.2. Then the resultant in the direction of D is $4.0-1.5-(2.5 \sin 30°) = 1.25$ N, and that in the direction of C is $2.0-1.0+(2.5 \cos 30°) = 3.16$ N. Figure 2.8 or Fig. 2.9 shows the final resultant. From Pythagoras' theorem the magnitude of the resultant $= \sqrt{(1.25^2+3.16^2)} = 3.4$ N. The inclination of the resultant is $\tan^{-1}(1.25/3.16) = 21\frac{1}{2}°$, confirming the less accurate graphical methods of (a), (b), and (c).

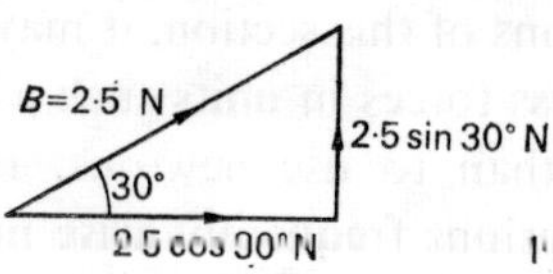

FIG. 2.7. Resolved components of force B of ex. 2.1.

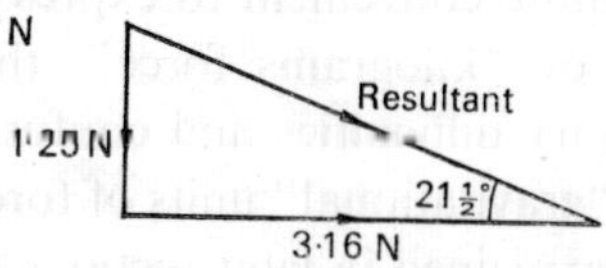

FIG. 2.8. Resultant by triangle (ex. 2.1).

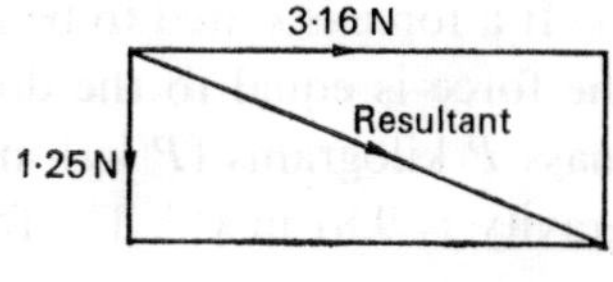

FIG. 2.9. Resultant by parallelogram (ex. 2.1).

2.2. Worked Example

A casting of weight 50 kgf is suspended from two ropes, one making an angle to the vertical of 30° and the other 45°. Calculate the tension in each rope (Fig. 2.10).

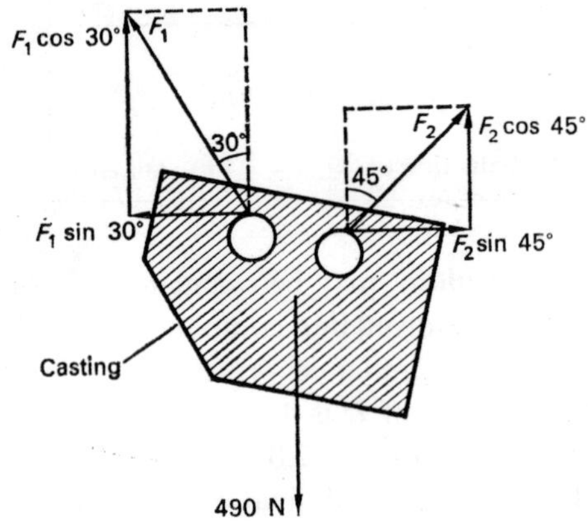

Fig. 2.10. Ex. 2.2.

Introduction

In this and some other questions of this section, it may be considered more convenient to express forces in units such as "grams weight" or "kilograms force" than to use newtons. However, unnecessary difficulties and confusions frequently arise in the use of such "gravitational" units of force in many of the more complex matters explained in later sections. It is therefore advisable almost invariably to avoid gravitational units of force by a simple conversion to newtons.

If a force is stated to be P kilograms force (P kgf) it means that the force is equal to the downward force of weight of a body of mass P kilograms (P kg) in a situation where the acceleration of gravity is 9.81 m s^{-2}. The force in newtons may then be calculated by using eqn. (3.2):

$$\text{force} = \text{mass} \times \text{acceleration}.$$

Thus P kgf expressed in newtons $= (P \text{ kg}) \times (9.81 \text{ m s}^{-2})$

$$= 9.81P \text{ newtons}. \qquad (2.1)$$

Solution

In the present case, where $P = 50$, the weight is $9.81 \times 50 = 490$ N.

Now if the casting is to be held in equilibrium, the two vertical components of force due to the ropes must together equal the weight, and the horizontal components must equal each other. Thus from the triangles or parallelograms of force shown in Fig. 2.10, for the vertical direction

$$F_1 \cos 30° + F_2 \cos 45° = 490 \qquad (2.2)$$

and for the horizontal direction

$$F_1 \sin 30° = F_2 \sin 45°. \qquad (2.3)$$

Then, using trigonometrical tables,

eqn. (2.2) gives $\quad 0.866F_1 + 0.707F_2 = 490$

and eqn. (2.3) gives $\quad 0.500F_1 - 0.707F_2 = 0.$

Adding these equations, $1.366F_1 = 490,$

$$\text{tension } F_1 = 360 \text{ N},$$

and $\quad \text{tension } F_2 = (0.500/0.707) \times 360 = 250 \text{ N}.$

2.3. The tensions in five of the six spokes of a wheel are 10, 80, 30, 40, and 50 N, taken in cyclic order. Determine the tension in the sixth spoke (a) by a six-sided polygon of forces, and (b) by some other method. [60 N.]

2.4. A lamp of weight 2.0 kgf is suspended from the ceiling by a light chain. A cord, attached to one of the links of the chain, is drawn so that the upper part of the chain makes an angle θ to the vertical when the cord is horizontal. Determine the angle θ (a) when the tension in the cord is twice the weight of the lamp, and (b) when the tension in the upper part of the chain is twice the weight of the lamp. [(a) 63°; (b) 60°.]

2.5. A jib, pivoted at the base of a wall, is held at an angle to the vertical by a horizontal tie, 3.0 m long between the upper end of the jib and a point on the wall 5.0 m vertically above the base of the jib. Calculate the tension in the tie and the thrust in the jib when the latter supports a load of 3.0 tonne at its upper end. [1.8×10^4 N; 3.4×10^4 N.]

2.6. Determine the additional force required for equilibrium with the following forces acting horizontally at a point, the directions being stated by compass bearings: 5.0 kgf, E 75° S; 8.0 kgf, E 50° S; 7.0 kgf, E 40° N; 10.0 kgf, W 10° S. [8.4 kgf; S 76° E.]

2.7. A heavy cable running above ground is supported on vertical poles. The route changes direction by 30° at one pole, where the side strain is taken symmetrically by a stay at 70° to the horizontal, attached to the top of the pole. Each part of the cable leaves the pole at 10° to the horizontal and has a tension of 250 N. Calculate the tension in the stay, assuming no horizontal force due to the pole itself. [370 N.]

Centre of Gravity

2.8. Worked Example

A square plate of metal of uniform thickness has a side of length 30 cm. One corner is cut off as a triangular piece having two sides of 20 cm each. Determine the position of the centre of gravity of the remainder given that the centre of gravity of a triangular plate is on a median at a point two-thirds of the median length from the vertex (Fig. 2.11).

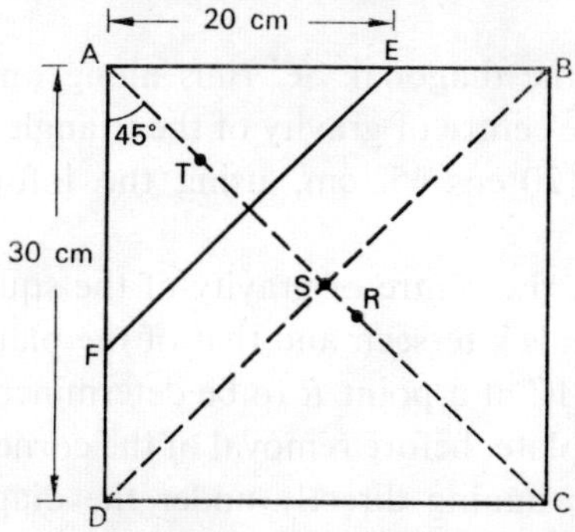

FIG. 2.11. Ex. 2.8.

Introduction

If an object is balanced on a knife edge, as in Fig. 2.12, the total turning moment of all parts of the weight on the right-hand side of the support must equal the total on the left-hand side. The object would still be balanced if its total weight acted as one downward force exactly through the support. If the object is turned

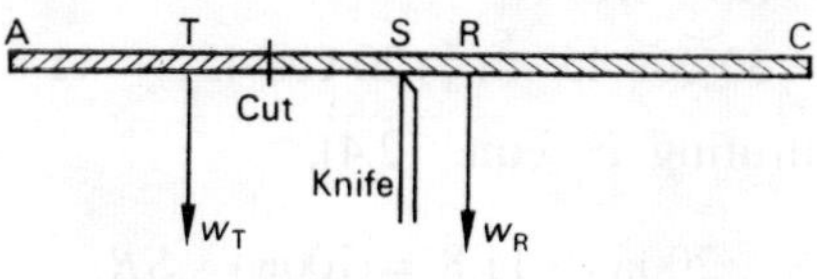

FIG. 2.12. Balanced plate of ex. 2.8.

over or round, a number of lines can be found, each showing the line of action of the total weight. All these lines pass through a common point called the centre of gravity.

The centre of gravity is then the point at which the whole weight of an object may be considered to act, however the object is placed.

Solution

In Fig. 2.11 the diagonal AC runs along one median of the triangle AEF. The centre of gravity of the triangle is therefore at T, where $AT = \frac{2}{3} \times 20 \cos 45°$ cm, using the information given in the question.

By symmetry, the centre of gravity of the square plate is at S, where the diagonals intersect, and that of the plate $EBCDF$ is also on the diagonal AC at a point R to be determined.

Consider the plate, before removal of the corner, to be balanced on a knife edge running directly under the diagonal BSD, as in Fig. 2.12 which shows a section along the diagonal ASC. The anticlockwise turning moment of the weight w_T, of the triangular portion, must equal the clockwise turning moment of the weight w_R, of the remainder of the plate.

$$\therefore \; w_T \times TS = w_R \times SR. \tag{2.4}$$

Now if w_1 = weight per square centimetre of the plate, then

$$w_T = w_1 \times 20^2/2 = 200 w_1$$

and $$w_R = w_1 \times (30^2 - 20^2/2) = 700 w_1.$$

But $$TS = AS - AT$$

$$= 30 \cos 45° - (\tfrac{2}{3})\, 20 \cos 45° = 11.8 \text{ cm}.$$

Then, substituting in eqn. (2.4),

$$200 w_1 \times 11.8 = 700 w_1 \times SR.$$

$$\therefore \; SR = 2360/700 = 3.4 \text{ cm}.$$

So the centre of gravity of the plate after the corner has been removed is 3.4 cm from the centre of the square on the line of symmetry.

An alternative method is to apply the principle of moments to the case of a knife edge under the corner A, running parallel to DB (Fig. 2.13). If a force F equal to the weight of the complete square w_S is applied upward under the centre of the square, the plate will be supported. Then the clockwise moments of w_T and w_R balance the anticlockwise moment of F.

$$\therefore\ (w_T \times AT)+(w_R \times AR) = F \times AS.$$

$$\therefore\ (200w_1 40/3\sqrt{2})+(700w_1 AR) = 900w_1 30/\sqrt{2}.$$

$$\therefore\ AR = (27000-8000/3)/700\sqrt{2} = 24.6 \text{ cm}.$$

Then $\quad SR = AR-AS = 24.6-30/\sqrt{2} = 3.4$ cm.

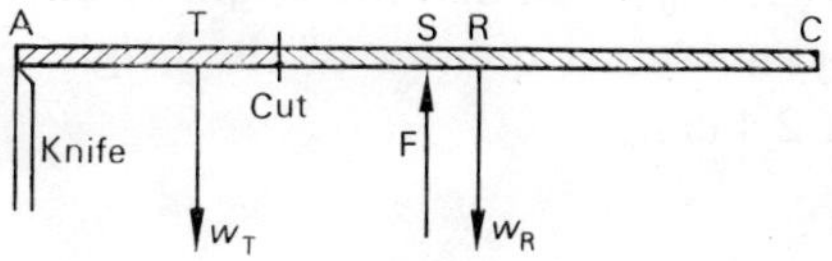

FIG. 2.13. Balanced plate of ex. 2.8.

2.9. A square flat plate of uniform thickness, 30×30 cm, has a square, 20×20 cm, cut from one corner so as to leave an *L*-shaped plate. Calculate the position of the centre of gravity. [11 cm from each long side.]

2.10. An equilateral triangle of thin wire, 10 cm along a side, has a mass of 20 g. How far can one vertex project over the edge of a table before overbalancing when a load of 5.0 g hangs from the vertex? [4.6 cm.]

2.11. A rectangular tray, 36×54 cm, is loaded unsymmetrically. The force required to lift one end alone is 32 N; the other 40 N. Calculate (a) the weight of the tray, and (b) the position of the centre of gravity. [(a) 72 N; (b) 24 cm from one side and from one end.]

2.12. A triangular sheet of metal, 40×40×30 cm, and of uniform thickness, is cut in half along a line parallel to the 30 cm side. Calculate the position of the centre of gravity of the four-sided piece given that the centre of gravity of a triangle is at the intersection of the medians. [5 cm from the 30 cm side.]

2.13. A metal cover of uniform thickness consists of a hollow cylinder of $\frac{1}{2}$ m length and $\frac{1}{2}$ m diameter, open at one end, and closed at the other by a hemispherical cap. Calculate the position of the centre of gravity given that the centre of gravity of the hemisphere is halfway along the radius which forms the axis of symmetry. [$\frac{3}{8}$ m from the open end.]

2.14. A right solid pyramid standing on a square base, 4.0×4.0 cm, is truncated so that the upper surface is square, 2.0×2.0 cm, at a vertical height of 6.0 cm from the base. Calculate the position of the centre of gravity given that the height of the centre of gravity of complete pyramid is a quarter of the height of the pyramid. [At a height of 2.4 cm.]

Inclined Plane

2.15. Worked Example

A parcel of mass 12 kg is held at rest on a smooth ramp, inclined at 50° to the horizontal, by a cord making an angle of 20° with the ramp. Calculate (a) the tension in the cord, and (b) the reaction of the ramp on the parcel (Fig. 2.14).

Introduction

A "smooth" surface is assumed to provide negligible friction. Then in the question, the parcel presses on the ramp and the ramp presses back on the parcel with forces perpendicular or "normal" to the surfaces. Thus the only force of the ramp on the parcel is the normal reaction F_n perpendicular to the ramp, as shown in Fig. 2.14. Since the behaviour of the parcel depends solely on forces acting on it and not on forces produced by it, the forces shown in Figs. 2.14 and 2.15 are only those on the parcel.

To calculate one of the unknown forces, tension F_t or reaction F_n, it is convenient to resolve forces into the directions perpendicular to the other force. Thus to calculate F_t the parallelograms in Fig. 2.14 are chosen to show components of tension and weight perpendicular to F_n. Parallelograms of force rather than single

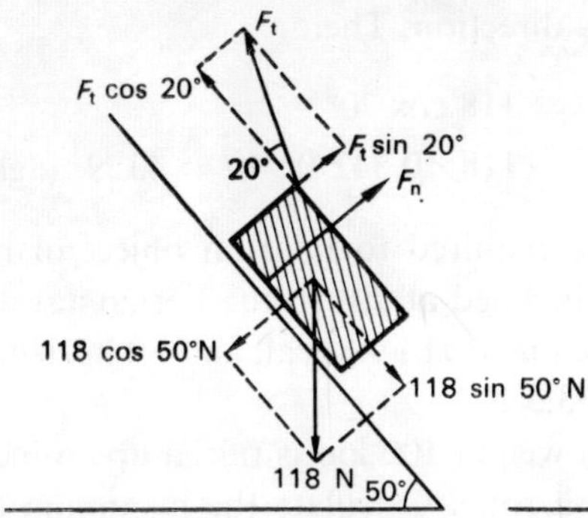

FIG. 2.14. Resolution of forces (ex. 2.15).

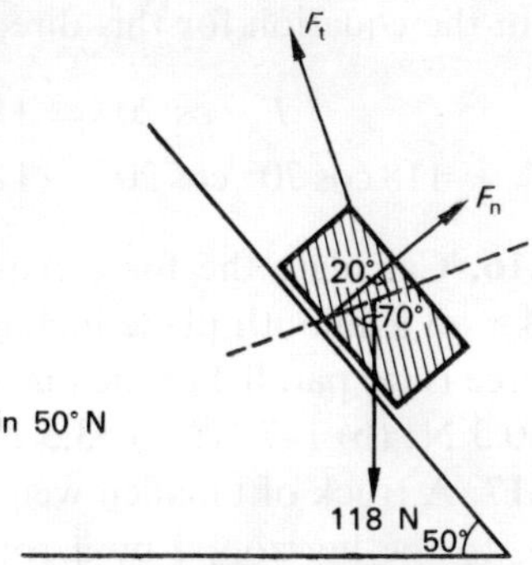

FIG. 2.15. Alternative direction of resolution (ex. 2.15).

triangles of force are shown so that the components appear to spring from the point of application of the tension and from the centre of gravity respectively.

Solution

In newtons, the weight of the parcel = 12×9.81 = 118 N.

(a) In the direction up and down the ramp, only two components have any effect: F_t cos 20° must be equal and opposite to 118 sin 50° N for equilibrium.

$$\therefore\ F_t = 118 \sin 50°/\cos 20° = 118\times 0.766/0.940 = 96 \text{ N}.$$

(b) Using the value of F_t just found, F_n may be calculated since the components of tension and weight perpendicular to the ramp surface combine with F_n to balance in that direction.

$$\therefore\ F_t \sin 20° + F_n = 118 \cos 50°.$$

$$\therefore\ F_n = 118 \cos 50° - 96 \sin 20° = (118\times 0.643) - (96\times 0.342)$$
$$= 76 - 33 = 43 \text{ N}.$$

Alternatively, the forces may be resolved into the direction perpendicular to F_t, shown in Fig. 2.15, so that F_t has no component in the equation for this direction. Then

$$F_n \cos 20° = 118 \cos 70°.$$

$$\therefore \quad F_n = 118 \cos 70°/\cos 20° = 118 \times 0.342/0.940 = 43 \text{ N} \quad \text{(again)}.$$

2.16. Calculate the force required to hold an object of mass 15.0 kg on a smooth plane inclined at 20° to the horizontal when the force is (a) parallel to the plane, (b) vertical, and (c) horizontal. [(a) 50.3 N; (b) 147 N; (c) 53.5 N.]

2.17. A truck of unladen weight 100 kgf is raised up an incline of 25° to the horizontal by a rope. Calculate the maximum load that may be raised in the truck if the force in the rope may not exceed (a) the unladen weight, and (b) the weight of the load. [(a) 136 kgf; (b) 73 kgf.]

2.18. A box of mass 40 kg is to be held at rest on a smooth plane inclined at 30° to the horizontal. A force of 100 N is applied by a rope inclined upward at 20° to the plane. Calculate the additional force required in a horizontal direction and the resulting reaction of the plane. [120 N; 360 N.]

Friction

2.19. Worked Example

A cord, inclined upward at 30° to the horizontal, is attached to a block of wood of mass 2.0 kg, resting on a horizontal surface with coefficient of friction 0.25. Calculate the minimum force in the cord needed to shift the block. If the surface is then tilted until it is parallel to the cord, calculate the minimum force needed (a) to hold the block at rest, and (b) to raise the block. In each case assume that the block remains flat on the surface.

Introduction

Whenever there is a tendency for sliding between two surfaces in contact, a pair of equal and opposite forces of friction arise.

These forces are tangential to the surfaces, and act one on each surface. Thus if the block in Fig. 2.16 is pushed or pulled to the right, the friction forces F_{fb} and F_{fs} come into play in addition to the normal reactions F_{nb} and F_{ns}, and act against any sliding. F_{fb} and F_{nb} act on the surface of the block, F_{fs} and F_{ns} on the supporting surface.

If the applied force F_a is increased from zero, both forces F_f increase, keeping $F_f = F_a$, in Fig. 2.16, until F_f reaches a limit at which sliding starts. This value $F_{f\,max}$ depends on the materials

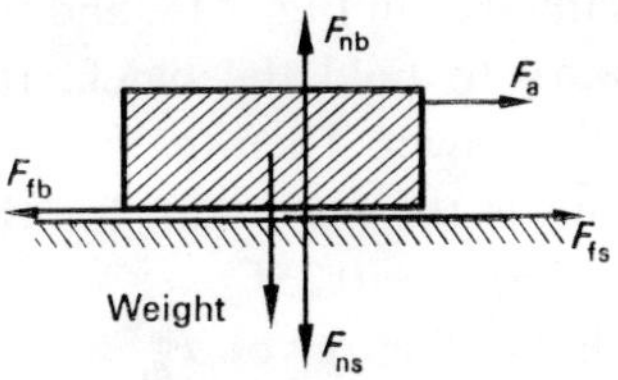

FIG. 2.16. Friction and normal reaction forces.

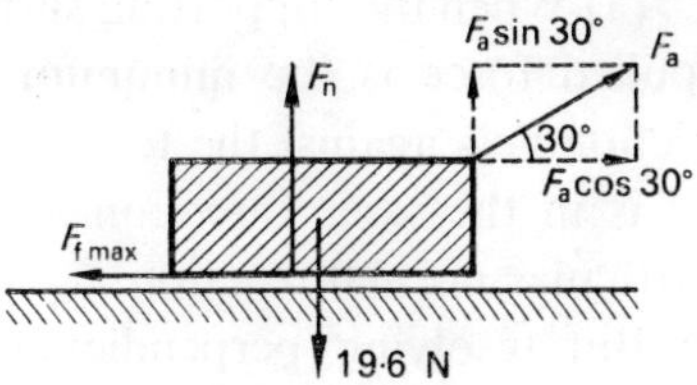

FIG. 2.17. Horizontal position (ex. 2.19).

and conditions of the surfaces and on the force of contact F_n (i.e. F_{nb} or F_{ns} since they are equal).

$$F_{f\,max} = \mu F_n, \qquad (2.5)$$

where μ is the coefficient of limiting or static friction and depends on the materials and conditions of the surfaces.

Solution

When the block is just about to move, due to a minimum applied force F_a as in Fig. 2.17, the block is in equilibrium. Therefore, resolving horizontally,

$$F_a \cos 30° = F_{f\,max} = \mu F_n. \qquad (2.6)$$

Also, in the vertical direction the upward forces must equal the weight $= 2 \times 9.81 = 19.6$ N.

$$\therefore \; F_n + F_a \sin 30° = 19.6. \qquad (2.7)$$

Note that the reaction F_n is less than the weight because of the upward component of F_a.

From eqn. (2.6),

$$0.866F_a = 0.25F_n. \quad \therefore F_n = 3.46F_a.$$

Then, substituting in eqn. (2.7),

$$3.46F_a + 0.50F_a = 19.6,$$

$$\therefore F_a = 19.6/3.96 = 5.0 \text{ N}.$$

(a) When the supporting surface is tilted as in Fig. 2.18, and the applied force is the minimum necessary to hold the block, the friction acts against the tendency to slide down. Then the friction F_{f_1} is in the same direction as F_{a_1}, and is the maximum for the particular normal reaction F_{n_1}. So $F_{f_1} = \mu F_{n_1} = 0.25F_{n_1}$.

But resolving perpendicular to the surface gives $F_{n_1} = 19.6 \cos 30°$.

$$\therefore F_{f_1} = 0.25 \times 19.6 \cos 30° = 4.25 \text{ N}.$$

Resolving up and down the surface gives $F_{a_1} + F_{f_1} = 19.6 \sin 30°$.

$$\therefore F_{a_1} = (19.6 \times 0.50) - 4.25 = 5.6 \text{ N}.$$

(b) If F_a is now increased, the friction first decreases, reaching zero when F_a alone balances the component of weight down the slope. Further increase of F_a produces a tendency to slide up, with friction opposing again. Eventually if F_a reaches the minimum value F_{a_2} needed for upward movement.

$$F_{a_2} = F_{f_2} + 1.96 \sin 30° = F_{f_2} + 9.8 \quad \text{(Fig. 2.19)}.$$

Here F_{f_2} is equal to the maximum value calculated for (a) but in the reverse direction.

$$\therefore F_{a_2} = 4.25 + 9.8 = 14 \text{ N}.$$

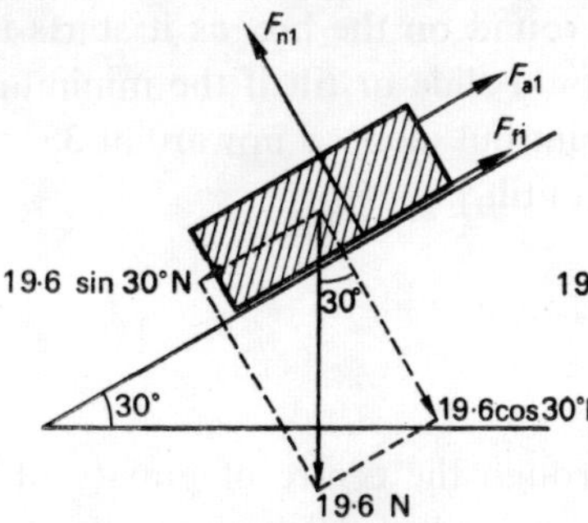

FIG. 2.18. Inclined position (ex. 2.19(a)).

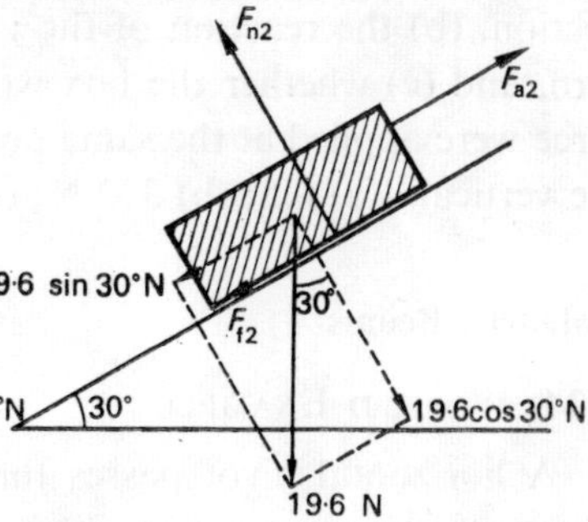

FIG. 2.19. Inclined position (ex. 2.19(b)).

2.20. A metal block rests on a board 45 cm long. To what height may one end of the board be raised before the block slips? The coefficient of friction = 0.60. [23 cm.]

2.21. A metal casting of mass 8.0 kg rests on a horizontal surface. A force of 30 N in a cord applied at 35° upward from the horizontal just moves the casting. Calculate the maximum friction force.

If the surface is then tilted and no force applied by the cord, at what angle will the block slip? [25 N; 22°.]

2.22. Each wheel of a motorized truck has a diameter of 10 cm and supports a load of 50 kg. One pair of wheels is driven through an axle by a torque of 15 Nm. Calculate the coefficient of friction between the wheels and the ground if the wheels just start to roll without slipping. What is the maximum tractive effort? [0.31; 300 N.]

2.23. A crate, in the shape of a cube, has a mass of 30 kg with centre of gravity at the centre of the cube. It rests on horizontal ground with a horizontal force applied at the centre of one upper edge in a direction perpendicular to the corresponding vertical face. The force is then made just sufficient to move the crate, which turns over just without sliding. Calculate (a) the coefficient of friction, (b) the reaction of the ground on the box as it starts to turn, and (c) whether the box would slide or tilt if the minimum force were applied at the same point but inclined upward at 35° to the vertical. [(a) 0.5; (b) 330 N; (c) tilt.]

Balanced Beams

2.24. Worked Example

A horizontal pivot passes through the centre of gravity of a horizontal beam weighing 20 N. A load of 3.0 kg hangs from a point 12.0 cm from the pivot, and a force of 50 N is applied at an angle θ to the beam at a point 8.0 cm from the pivot. Determine the angle θ for equilibrium and the two values of reaction at the pivot, assuming the direction of the applied force to be (a) as in Fig. 2.20, and (b) as in Fig. 2.22.

Solution

(a) The horizontal component of F_a in Fig. 2.20 has no turning moment since it passes through the pivot, but the vertical component $50 \sin\theta$ has a clockwise turning moment $50 \sin\theta \times 8.0 \times 10^{-2}$ Nm about the pivot P. For balance, this must equal the anticlockwise moment of the load force.

$$\therefore\ 50 \sin\theta \times 8.0 \times 10^{-2} = 3.0 \times 9.81 \times 12 \times 10^{-2}.$$

$$\therefore\ \sin\theta = 3.0 \times 9.81 \times 12/400 = 0.884 \qquad \therefore\ \theta = 62°.$$

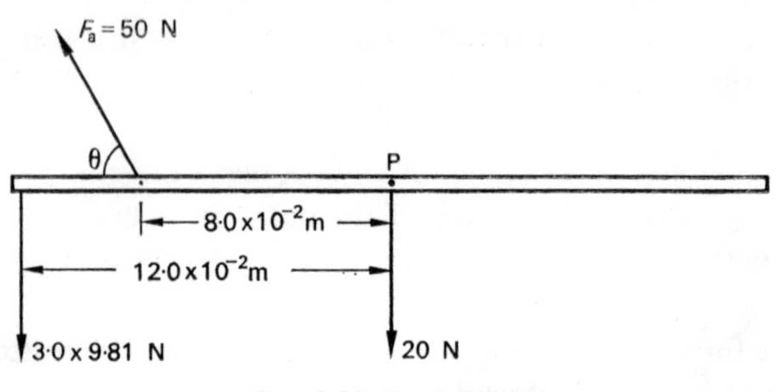

Fig. 2.20. Ex. 2.24(a).

Although the moments of these forces are balanced, the forces themselves are unequal; the vertical component of F_a exceeds the load force. So the pivot P receives a thrust equal to their difference in addition to a thrust due to the weight of the beam. Therefore the total downward thrust on the pivot in Fig. 2.20 is equal to

$$20+(3.0\times9.81)-(50\times0.884) = 20+29.4-44.2 = 5.2 \text{ N}.$$

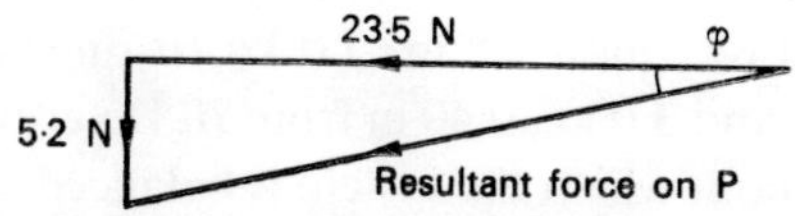

FIG. 2.21. Resultant thrust on the pivot (ex. 2.24(a)).

The horizontal component of F_a causes a horizontal thrust on the pivot

$$= 50 \cos 62° = 50\times0.469 = 23.5 \text{ N}.$$

Then the resultant thrust on the pivot is found by Pythagoras' theorem applied to Fig. 2.21.

Therefore resultant thrust on $P = \sqrt{(23.5^2+5.2^2)} = 24$ N at angle φ, where $\varphi = \tan^{-1} 5.2/23.5 = 12\frac{1}{2}°$.

The reaction of the pivot on the beam is equal and opposite to the thrust on P.

(b) When the applied force is as in Fig. 2.22, the calculation would be the same for F_b as for F_a, but the horizontal component

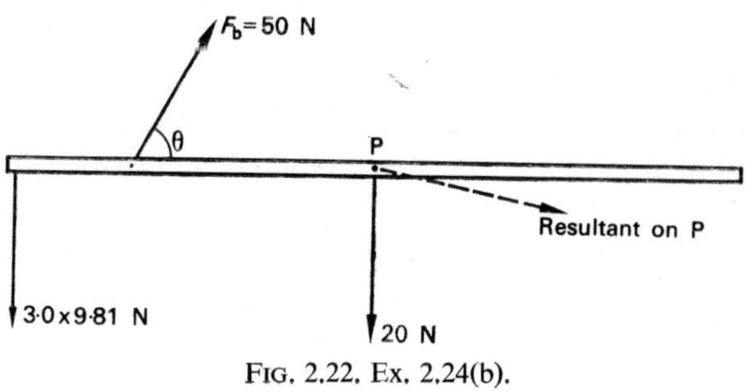

FIG. 2.22. Ex. 2.24(b).

of 23.5 N is reversed. Then the reaction of the pivot is again 24 N at an angle of $12\frac{1}{2}°$, but in a different direction from (a).

2.25. A uniform beam of length 2.0 m and weight 50 N is carried by two men, one at each end. A load of 51 kg hangs on the beam at a point between the men. Calculate the position of the load if one man, *A*, supports a force of 300 N. [0.9 m from *A*.]

2.26. Three loads hang at points on a uniform beam of negligible weight and of length 1.5 m: 7.0 kg, at one end *A*; 6.0 kg at the other end *B*; and 3.0 kg, 0.40 m from *B*. Determine the position of the fulcrum about which the system is balanced. [77 cm from *A*.]

2.27. A uniform ladder, 5.0 m long, weighing 15 kgf, rests against a smooth wall with its base 3.0 m from the wall. Find (a) the forces at the ends of the ladder when a man of 75 kg is halfway up, and (b) the vertical height to which a man of 75 kg can climb before the ladder slips. Assume that the coefficient of limiting friction with the ground is 0.50 and that the ground is horizontal and the wall vertical. [(a) 940 N, 330 N; (b) 2.8 m.]

2.28. A metal plate of uniform thickness is shaped as an equilateral triangle and weighs 2.0 kgf. It hangs on a cord attached to one corner with loads of 4.0 kg and 12.0 kg attached one each to the other corners. Calculate the angle to the horizontal made by the base of the triangle. [15°.]

2.29. A chemical balance has a beam of mass 100 g with arms of length 10 cm each. The pans are of mass 75 g each, including hook and cradle. When the beam is horizontal its centre of gravity is 0.10 mm vertically below the mid-point of the line joining the knife edges from which the pans are suspended, and the central knife edge is 0.10 mm vertically above the mid-point. Consider the weighing of an object of mass (a) 200 g, and (b) 2.0 g. In each case calculate the sensitivity θ/m, where θ = the angle of deflection caused by a slight difference m between the masses of the loads in the two pans. [(a) 1.3 rad g^{-1}; (b) 2.8 rad g^{-1}.]

Shearing Force and Bending Moment

2.30. Worked Example

A light beam, 1.2 m long, is supported horizontally at one end and at the centre. A load of 2.0 kg is attached at the free end and a load of 10.0 kg at the point midway between the supports. Calculate the shearing force and bending moment at points 20 cm to each side of the centre, and draw complete shearing force and bending moment diagrams (Fig. 2.23).

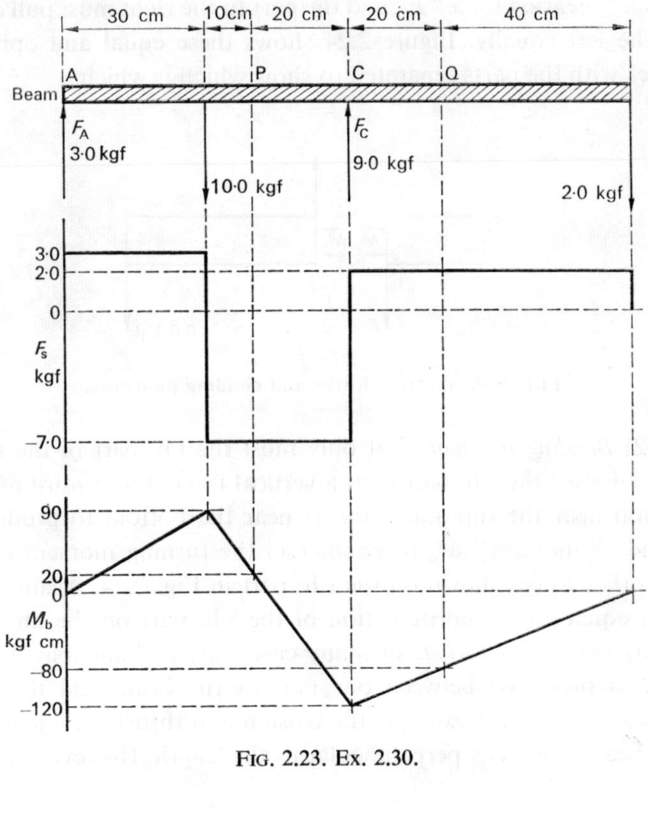

FIG. 2.23. Ex. 2.30.

Introduction

When a beam is fixed or supported at one or more places, and external forces are applied at other places, the beam tends to shear and to bend. In cases such as that of Fig. 2.23, where all forces may be shown in one plane, the resulting forces within the beam may be dealt with under three headings.

For example, consider a section through Q in Fig. 2.23:

(1) *Shearing force.* If the beam were cut through at Q, the part to the right would fall due to the load at the end. Therefore when the beam is not cut, the part to the left must support the right with a "shearing force" F_s, and the part to the right must pull down on the left equally. Figure 2.24 shows these equal and opposite forces with the parts separated to show which is which.

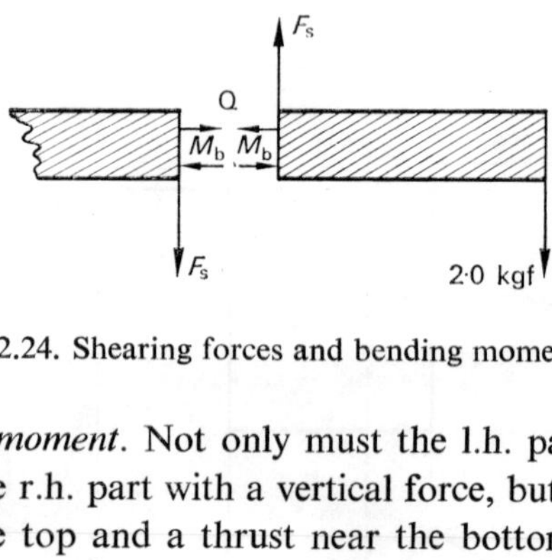

FIG. 2.24. Shearing forces and bending moments.

(2) *Bending moment.* Not only must the l.h. part of the beam at Q support the r.h. part with a vertical force, but it must exert a tension near the top and a thrust near the bottom to produce a "bending moment" M_b to counteract the turning moment of the two other forces shown on the r.h. part in Fig. 2.24. Again, there is an equal and opposite action of the r.h. part on the l.h. part.

(3) *Tension or thrust.* In many cases no resultant longitudinal force is produced between one part of the beam and the next. Thus in the present example the tension and thrust are equal and the shearing force is perpendicular to the length. However, if there

were longitudinal components of external force, a resultant tension or thrust would be produced.

If the force at the free end had been upward, the directions of F_s and M_b, shown in Fig. 2.24, would have been reversed. It is common to reckon F_s and M_b to be positive when upward and clockwise respectively on the r.h. part. Thus in Fig. 2.24, F_s is positive but M_b is negative. This convention makes the shearing force positive where the bending moment diagram has a positive upward gradient; in fact F_s equals the gradient of M_b as seen in Fig. 2.23.

The values of F_s and M_b at P and other parts of the beam may be found if the reactions at the supports A and C in Fig. 2.23 are determined as in detail below.

Solution

The reactions of the supports are found by taking moments first about one support and then about the other. Thus equating moments about A in Fig. 2.23 gives

$$(10.0\times 30)+(2.0\times 120) = F_C\times 60,$$

where F_C is the reaction of the support on the beam at the centre C.

Then

$$F_C = (300+240)/60 = 9.0 \text{ kgf}.$$

Similarly, for moments about C,

$$(F_A\times 60)+(2.0\times 60) = 10.0\times 30.$$

$$\therefore\ F_A = (300-120)/60 = 3.0 \text{ kgf}.$$

F_s can now be found at any point by calculating either the upward resultant of all external forces on the l.h. part or downward on the r.h. part. Thus, starting from end A,

$$F_s = 3.0 \text{ kgf} = (10.0-9.0+2.0) \text{ kgf}$$

for all points up to that where the 10.0 kg load hangs. Beyond this and up to C the forces on the l.h. part are 3.0 kgf and 10.0 kgf, making $F_s = +3.0-10.0 = -7.0$ kgf, and beyond C, $F_s = +2.0$ kgf.

At A there is no bending moment because there is nothing on the left to produce it. Then M_b increases linearly and positively from zero to (3.0×30) kgf cm, increasing because it equals only the clockwise turning moment of F_A which increases in proportion to distance from A.

Beyond the point where the load of 10.0 kg hangs, M_b is decreased by the anticlockwise moment of this load, reaching a value at C of

$$(3.0\times60)-(10.0\times30) = 180-300 = -120 \text{ kgf cm}.$$

Before this, at P,

$$M_b = (3.0\times40)-(10.0\times10) = 120-100 = 20 \text{ kgf cm}.$$

Then M_b returns toward zero due to the increasing clockwise moment of F_C. At Q,

$$M_b = (3.0\times80)-(10.0\times50)+(9.0\times20) = 240-500+180$$
$$= -80 \text{ kgf cm},$$

or, more easily by considering the single force on the right of Q,

$$M_b = -2.0\times40 = -80 \text{ kgf cm}.$$

2.31. A light horizontal cantilever fixed in and projecting from a wall at one end carries three loads, 4.0, 5.0, and 6.0 kg at distances from the fixed end of 100, 150, and 200 cm respectively. Calculate the shearing force and bending moment at the fixed end and at distances of 125 and 175 cm from the fixed end. [15, 11, 6 kgf; −2350, −575, −150 kgf cm.]

2.32. A light beam, 2.4 m long, is supported horizontally at both ends and is loaded at points $\frac{1}{4}$, $\frac{1}{2}$, and $\frac{3}{4}$ of the length from one end with loads of 10, 15, and 20 kg respectively. Calculate the shearing force and bending moment at points one-third and two-thirds of the length from one end. [+10, −5 kgf; +140, +160 kgf m.]

2.33. A uniform heavy beam, 3.0 m long and 50 kgf in weight, is fixed at one end as a horizontal cantilever and carries a load of 75 kg at the free end. Calculate the shearing force and bending moment at the fixed end (a) when no other support is provided, and (b) when an additional support provides an upthrust of 40 kgf at a point 2.0 m from the fixed end. [(a) 125 kgf, −300 kgf m; (b) 85 kgf, −220 kgf m.]

CHAPTER 3

FORCE AND ACCELERATION

Horizontal or Vertical Motion

3.1. Worked Example

A body of mass 8.0 kg rests on a horizontal table. Calculate the force required to give the body a horizontal velocity of 5.0 m s^{-1} in a time of 0.50 s assuming (a) a perfectly smooth surface, and (b) an opposing force of 40 N due to friction.

Introduction

To change the velocity of a body a resultant force must be applied to it in the direction of the change, i.e. in the direction of the acceleration. The magnitude of the acceleration a depends upon the magnitude of the resultant force F and upon the property of the body called its mass m. The relationship between these quantities is given by Newton's second law in the form

$$F \propto ma. \tag{3.1}$$

This may be written as an equation,

$$F = ma, \tag{3.2}$$

if the units of F, m, and a are such that $F = 1$ when $m = 1$ and $a = 1$. Thus in the metre–kilogram–second (M.K.S.) system, $F = 1$ newton (N) when $m = 1$ kilogram (kg) and $a = 1$ metre per second per second (m s^{-2}), i.e. 1 N is the force which gives a body of mass 1 kg an acceleration of 1 m s^{-2}. Various other systems of units suit eqn. (3.2), but their use is dying out and is not considered here.

Solution

(a) Assuming a perfectly smooth surface, there is no opposition to sliding, and assuming a perfectly horizontal supporting surface, no other support is required to oppose the weight of the body. So in (a) the force is required only to give horizontal acceleration

$$a = (v-u)/t = 5.0/0.50 = 10 \text{ m s}^{-2}.$$

Then, from eqn. (3.2),

$$F = ma = 8.0 \times 10 = 80 \text{ N}. \quad \text{(a)}$$

(b) The friction makes no difference to the mass of the body but introduces a force acting against the applied force. The applied force must then exceed the friction by the amount required for acceleration.

$$\therefore \text{ required force} = 40+80 = 120 \text{ N}. \quad \text{(b)}$$

3.2 Worked Example

Calculate the mass and true weight of an object which appears to weigh 2.50 kgf on a spring balance in a lift, which is accelerating upward at a rate of 0.50 m s^{-2}, where the acceleration of gravity is 9.81 m s^{-2}.

Introduction

Since a body normally requires a force to support it at rest, it is said to have a weight, i.e. a force tending to cause downward acceleration. This force varies with latitude and altitude, and depends upon any buoyancy of the surrounding medium and upon the movement of the body with the earth or relative to it. When the movement or the medium has a considerable effect, it is usual to distinguish between the apparent weight, under those conditions, and the true weight. The latter is the force, due to gravitation and rotation of the earth, which would cause acceleration from rest relative to the earth if the body were unsupported in a vacuum. Even this true weight is not a constant property of the body; it varies with latitude and altitude because it is due to the two effects of universal gravitation and centripetal acceleration (see Chapter 11). The variation of the true weight is shown by the variation of the resulting acceleration g, commonly called the acceleration of gravity.

The true weight may be calculated from the appropriate value of g and the mass m, using eqn. (3.2):

$$\text{force} = \text{mass} \times \text{acceleration}.$$

$$\therefore \text{ true weight} = \text{mass} \times \text{acceleration of gravity}$$

$$= mg. \tag{3.3}$$

Unless there are indications to the contrary g may be taken to be 9.81 m s^{-2}.

Weight, like other forces, is measured in newtons in the M.K.S. system. The unit kilogram-force (kgf) or kilogram-weight (kgwt) is best avoided because it does not give unit acceleration when acting on unit mass in the M.K.S. system, but gives the standard acceleration of gravity, 9.81 m s^{-2}.

Thus, using eqn. (3.2),

$$F = ma.$$

∴ 1 kgf, expressed in newtons,

$$= 1 \times 9.81 = 9.81 \text{ N}.$$

It follows that 1 N approximately equals the weight of a body of mass $\frac{1}{10}$ kg.

When an object hangs at rest on a spring balance, the extension of the spring depends upon the apparent weight of the object.

The balance is usually calibrated so that the extension indicates not the weight but the mass of the object whose true weight produces that extension when $g = 9.81\ \mathrm{m\,s^{-2}}$. However, the extension for a given mass may be increased or decreased by using the balance under abnormal conditions, as in the question. Although under such conditions the scale of the spring indicates an incorrect mass, it gives a measure of the force which the spring exerts by its tension. From this the mass and true weight of the object may be deduced if the conditions are known, as follows.

Solution

The indication of 2.50 kg shows that the tension in the spring is 2.50×9.81 N, and would normally support an object of this weight. Yet to produce upward acceleration, as opposed to supporting at rest, the tension must exceed the weight.

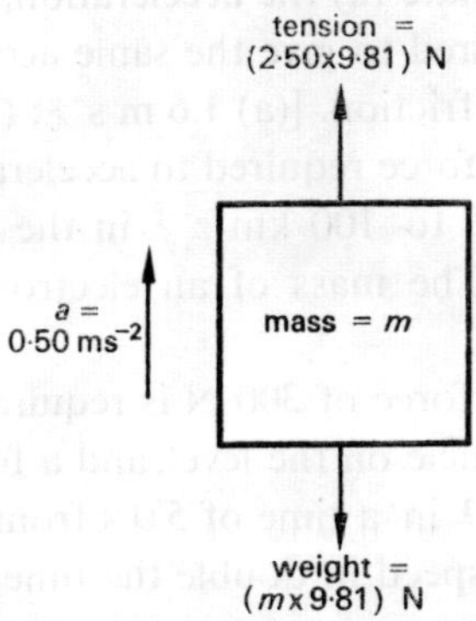

FIG. 3.1. Ex. 3.2.

Now the true weight $= m \times 9.81$ N, where m is the true mass.

Then the resultant force on the object = tension − weight (Fig. 3.1).

$$\therefore F = (2.50 \times 9.81) - (m \times 9.81).$$

But

$$F = ma = m \times 0.50.$$

$$\therefore \ (2.50 \times 9.81) - (m \times 9.81) = m \times 0.50.$$

$$\therefore \ m = 2.50 \times 9.81/10.31 = 2.378 = 2.38 \text{ kg} \quad \text{(3 sig. fig.)}.$$

Then the true weight

$$= m \times 9.81 = 2.378 \times 9.81 = 23.3 \text{ N} \quad \text{(3 sig. fig.)}.$$

3.3. A force of 110 N, acting on a mass of 4.0 kg, produces an acceleration of 20 m s^{-2}. Calculate the force of resistance. [30 N.]

3.4. A metal ball of mass 10 g falls in oil with an acceleration of 3.0 m s^{-2} at one instant. Calculate the opposing force due to the oil. [6.8×10^{-2} N.]

3.5. A crate of mass 50 kg slides on a horizontal floor under the action of an applied horizontal force of 200 N and against a friction force of 120 N. Calculate (a) the acceleration, and (b) the applied horizontal force required to give the same acceleration with twice the mass and half the friction. [(a) 1.6 m s^{-2}; (b) 220 N.]

3.6. Calculate the force required to accelerate an electron from a speed of 100 m s^{-1} to 100 km s^{-1} in the same direction in a distance of 5.0 cm. The mass of an electron is 9.1×10^{-31} kg. [9.1×10^{-20} N.]

3.7. A horizontal force of 300 N is required to maintain constant velocity of a vehicle on the level, and a force of 1100 N gives it a speed of 6.0 m s^{-1} in a time of 5.0 s from rest. What force is required for half the speed in double the time, assuming constant resistance? [500 N.]

3.8. A car of mass 750 kg is accelerated from rest by a constant force of 500 N for a distance of 200 m. Calculate the maximum speed reached and the brake force required to stop the car in a distance of 20 m. [16 m s^{-1}; 5000 N.]

3.9. It is reckoned that the driver of a vehicle with good brakes on a dry road may take a distance of 25 m to stop from a speed of

50 km h^{-1} in a sudden emergency. Of this distance, 10 m is covered during the driver's reaction time. Calculate (a) the reaction time, and (b) the resultant force on a driver of 75 kg if he remains in his seat and the brake force is constant during application. [(a) 0.72 s, (b) 480 N.]

3.10. A van of unladen mass $\frac{3}{4}$ tonne has a minimum braking time of 2.0 s in coming to rest from 48 km h^{-1}. When the same force is used, calculate the distance covered in coming to rest from a speed of 30 km h^{-1} (a) without, and (b) with a load of $\frac{1}{2}$ tonne. [(a) 5.2 m; (b) 8.7 m.]

Momentum

3.11. Worked Example

A gun of mass 50 tonne gives a shell of mass 500 kg a velocity of 800 m s^{-1}. Calculate (a) the impulse on the shell, and (b) the relative velocity of the shell and the gun as the shell leaves the muzzle. Assume that the retarding forces on the shell and the gun are negligible during the impulse.

Introduction

Although Newton's second law of motion has been quoted in the expression (3.1) and used as an equation (3.2) to define the unit of force, the primary statement of the law involves the concept of momentum, i.e. the product of mass and velocity. In terms of this, the law states that the rate of change of momentum of a body is proportional to the resultant force acting on the body in the same direction.

Now when an object receives an impulse, the resultant force generally varies during the time of application, but it is often sufficient to consider the mean value F to act for a time t. If, due to F, the velocity of the object of mass m is increased from u to v, the momentum is increased from mu to mv, and the law gives the relationship

$$(mv-mu)/t \propto F. \tag{3.4}$$

At this point, the explanation connects with ex. 3.1, since $(v-u)/t$ is the mean acceleration a, and so $ma \propto F$ for constant m. Then, using absolute units such as those of the M.K.S. system, (3.4) becomes an equation:

$$mv-mu = Ft, \tag{3.5}$$

where the product Ft is the impulse.

When two objects A and B are considered, Newton's third law states that the action of A on B is equal and opposite to the reaction of B on A. But the times during which the action and reaction operate on A and B are the same. Therefore the impulses on A and B are equal and opposite, and likewise the changes of momentum. Hence the principle of conservation of momentum:

> *When no external forces act on a system, the total momentum remains constant in any interaction between its parts, if velocities in one direction are reckoned positive and those in the opposite direction negative.* (3.6)

Solution

(a) In the case of the gun and shell, the shell starts from rest with initial momentum $m_s u_s = 0$. But the final momentum

$$m_s v_s = 500 \times 800 = 4 \times 10^5 \text{ kg m s}^{-1}.$$

Then, from eqn. (3.5), the impulse given to the shell

$$= Ft = m_s v_s - m_s u_s = 4 \times 10^5 - 0$$

$$= 4 \times 10^5 \text{ N s.} \tag{a}$$

(b) The impulse of the reaction on the gun is equal and opposite to that of the action on the shell and causes a reverse or negative momentum.

$$\therefore \text{ gain of momentum of gun} = -4 \times 10^5 \text{ kg m s}^{-1}$$

$$= m_g v_g - m_g u_g.$$

But the gun starts from rest.

$$\therefore \; u_g = 0.$$

$$\therefore \; m_g v_g = -4\times 10^5, \quad \text{with} \quad m_g = 50 \text{ tonne} = 50\times 10^3 \text{ kg}.$$

$\therefore$ recoil velocity of gun $v_g = -4\times 10^5/(50\times 10^3) = -8 \text{ m s}^{-1}$.

Then the relative velocity of shell and gun

$$= v_s - v_g = 800-(-8) = 808 \text{ m s}^{-1}. \qquad \text{(b)}$$

Alternatively, the above statement (3.6) may be used:

Total initial momentum, reckoned positive in the direction of the shell $= m_s u_s + m_g u_g = 0$, since both shell and gun are at rest. Now total initial momentum $=$ total final momentum.

$$\therefore \; 0 = m_s v_s + m_g v_g.$$

$\therefore \; v_g = -m_s v_s/m_g = -500\times 800/(50\times 10^3) = -8 \text{ m s}^{-1}$ (as before)

3.12. A truck of mass 25 tonne, travelling at a speed of 9 kmh^{-1}, collides with a second truck of mass 15 tonne, travelling at 4 km h^{-1} in the same direction. They part with a relative velocity of 3 km h^{-1}. Calculate the final velocity of each. [6 km h^{-1}; 9 km h^{-1}.]

3.13. A ball of mass 50 g falls from rest at a height of 1.0 m on to a horizontal surface and rebounds to a height of 0.50 m. Calculate the impulse on the surface. [0.38 N s.]

3.14. An α-particle, travelling at a speed of 1000 m s^{-1}, collides with a stationary hydrogen atom having a mass of a quarter of that of the α-particle. Assuming a perfectly elastic collision, calculate the final speed of the α-particle if it is undeflected. [600 m s^{-1}.]

3.15. A vehicle of mass 1000 kg is travelling horizontally at a velocity of 108 km h^{-1}. Brakes with a retarding force of 5000 N are applied for a distance of 50 m before they fail completely. Calculate the velocity after subsequent collision with a second vehicle of mass 1500 kg, travelling freely in the same direction at a velocity which is increased by the collision from 54 km h^{-1} to 63 km h^{-1}. [16 m s^{-1}.]

Motion on Gradient

3.16. Worked Example

An object on a ramp just starts to slide when the angle is increased to 25° to the horizontal. The object then slides 17 cm in a time of 0.75 s from rest. Calculate the coefficients of static and of dynamic friction.

Solution

When the object is just about to slide, the friction force F_{f_1} is a maximum equal to the opposite component of weight mg (Fig. 3.2).

$$\therefore\ F_{f_1} = mg \sin 25° = 0.423mg.$$

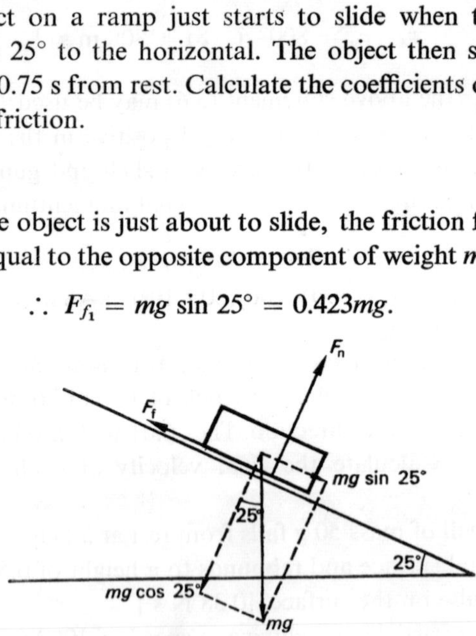

FIG. 3.2. Ex. 3.16.

The normal reaction

$$F_n = mg \cos 25° = 0.906mg.$$

$$\therefore \text{ coefficient of static friction} = F_{f_1}/F_n$$
$$= 0.423mg/(0.906mg) = 0.47.$$

As the object slides, the friction force drops below the maximum static value, and the resultant force on the object is

$$F = mg \sin 25° - F_{f_2} = (0.423 \times 9.81m) - F_{f_2}. \qquad (3.7)$$

Now the acceleration is given by the equation

$$s = ut + \tfrac{1}{2}at^2,$$

where $u = 0$ for sliding from rest, $s = 0.17$ m, $t = 0.75$ s.

$$a = 2s/t^2 = 2 \times 0.17/0.75^2 = 0.605 \text{ m s}^{-2}.$$

Then

$$F = ma = 0.605m.$$

Also $F_{f_2} = \mu_d F_n$, where μ_d is now the coefficient of dynamic friction, although $F_n = 0.906\ mg$ as before. $F_{f_2} = 0.906 \times 9.81\ m\mu_d$.

Then, substituting in eqn. (3.7),

$$0.605m = (0.423 \times 9.81m) - (0.906 \times 9.81m\mu_d).$$

$$\therefore\ 0.605 = 4.15 - 8.89\mu_d.$$

$$\therefore \text{ coefficient of dynamic friction, } \mu_d = 3.545/8.89 = 0.40.$$

3.17. Calculate the angle of a ramp on which a parcel of mass 2.0 kg would slide downward with an acceleration of 3.0 m s^{-2} against friction forces amounting to 5.0 N. [34°.]

3.18. A locomotive draws a train of mass 250 tonne with a constant coupling force of 5.0×10^4 N from rest up an incline of 1 in 200 for a distance of 500 m, and then along the level for a further 500 m. Calculate the speed at the top of the incline, the final speed, and the average speed. Neglect frictional resistance. [12, 19, 8.8 m s^{-1}.]

3.19. A crate of mass 30 kg slides horizontally at a speed of 3.0 m s^{-1} as it approaches a smooth upward ramp of gradient $\sin^{-1} 0.10$. Calculate how far it will travel up the ramp in a time of 1.0 s, when assisted by a force of 10 N. [2.7 m.]

3.20. A motor-cycle of mass 150 kg produces a driving force of 500 N. Resistance to motion is assumed to be a constant force of 50 N. Calculate the velocity reached in 50 m from rest (a) on the flat, and (b) up a gradient of $\sin^{-1} 0.10$. Also calculate (c) the gradient that would just prevent acceleration. [(a) 17 m s^{-1}; (b) 14 m s^{-1}; (c) $\sin^{-1} 0.31$.]

3.21. An object will just slide unassisted down a 30° slope. What would be its acceleration unassisted on a 45° slope with the same coefficient of friction? [2.9 m s^{-2}.]

3.22. A box slides unassisted from rest through a distance of 1.0 m down a 30° slope in a time of 1.6 s. Calculate the coefficient of friction. [0.48.]

Coupled Objects

3.23. Worked Example

A locomotive of mass 50 tonne hauls a train of mass 200 tonne up an incline of $\sin^{-1} 1/300$ with a coupling force of 6.0×10^4 N against a resistance of 1.0% of the weight of each part. Calculate (a) the acceleration, and (b) the driving force due to friction between driving wheels and rails.

Solution

In accordance with Newton's third law, two equal and opposite coupling forces F_C are exerted, one forward on the train and the other backward on the loco (Fig. 3.3). Also the resistances to motion are each given by $F_R = mg/100$, and the components of weight acting in the direction opposite to motion are each given by $mg/300$, since $\sin \theta = 1/300$.

Therefore the effects of resistance and weight together amount to

$$mg(1/100+1/300) = 4mg/300$$

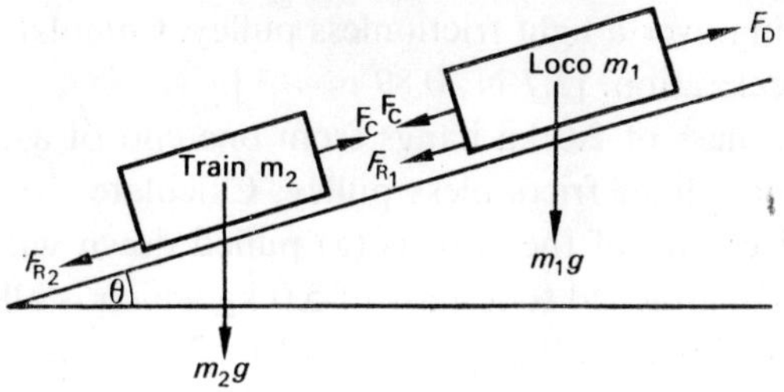

FIG. 3.3. Ex. 3.23.

Then, for the loco, resultant force

$$F_1 = F_D - F_C - (4m_1g/300), \qquad (3.8)$$

where F_D is the driving force, and m_1g is the weight of the loco.

For the train, resultant force $F_2 = F_C - (4m_2g/300)$, where m_2g is the weight of the train.

$$\therefore\ F_2 = (6.0\times10^4) - (4\times200\times10^3\times9.81/300) = 3.38\times10^4\ \text{N}.$$

$$\therefore\ \text{acceleration } a = F_2/m_2 = 3.38\times10^4/(200\times10^3)$$
$$= 0.17\ \text{m s}^{-2}. \qquad \text{(a)}$$

Then for the loco, $F_1 = m_1a = 50\times10^3\times0.17 = 8.5\times10^3$ N. Substituting in eqn. (3.8) and rearranging gives

$$\text{driving force } F_D = F_1 + F_C + (4m_1g/300)$$
$$= (8.5\times10^3) + (6.0\times10^4) + (4\times50\times10^3\times9.81/300)$$
$$= 7.5\times10^4\ \text{N}. \qquad \text{(b)}$$

3.24. A lift of mass 500 kg contains a man of mass 75 kg and has an acceleration of (1) 1.0 m s^{-2} upward, and (2) 1.0 m s^{-2} downward. Calculate (a) the force of the man on the floor, and (b) the force of the cable on the lift in each case. [(1a) 810 N; (1b) 6.2×10^3 N; (2a) 660 N; (2b) 5.1×10^3 N.]

3.25. Loads of mass 250 g and 300 g hang from the ends of a string passing over a light frictionless pulley. Calculate the tension and the acceleration. [2.7 N; 0.89 m s^{-2}.]

3.26. A mass of 2.0 kg hangs from one end of a cord passing up and over a light frictionless pulley. Calculate the acceleration when the free end of the cord is (a) pulled down with a force of 50 N, and (b) attached to a mass of 5.0 kg which is allowed to fall when the cord is taut. [(a) 15 m s^{-2}; (b) 4.2 m s^{-2}.]

3.27. A block of mass 5.0 kg rests on a horizontal table. It is pulled horizontally by a string which passes over a light frictionless pulley at the edge of the table and supports a second block of mass 2.0 kg which hangs at the end. Calculate the tension in the string and the acceleration when the system is released with a friction force of 15 N between the first block and the table. [18 N; 0.66 m s^{-2}.]

3.28. Two buckets, each of mass 3.0 kg, hang on opposite ends of a cord passing over a light pulley of diameter 30 cm, running on a shaft of diameter 2.0 cm. The friction force between the wheel and shaft is 150 N, and the inertia of the wheel may be neglected. Calculate the force required on one empty bucket (a) to raise, and (b) to lower a load of 25 kg in the other bucket, with an acceleration of 0.50 m s^{-2}. [(a) 270 N; (b) 220 N.]

3.29. A lorry of mass 2000 kg leads a trailer of mass 1000 kg down a hill of gradient $\sin^{-1}\frac{1}{7}$ at a speed of 30 km h^{-1}. The brakes work only on the lorry. Calculate the forces of the tow-bar and brakes to bring the vehicles to rest in a distance of 100 m. [1.7×10^3 N; 5.2×10^3 N.]

CHAPTER 4

ENERGY AND POWER

Work and Energy

4.1. Worked Example

A load of 60 kg is lifted freely against gravity by a constant upward vertical force of 1000 N. Calculate the potential energy acquired in a period of 0.40 s starting from rest.

Introduction

Energy exists in various forms, such as potential energy, which is associated with an object which may fall from a height, and kinetic energy, which is associated with an object which is moving. Energy cannot be created or destroyed, but may be converted from one form to another, or transferred from one situation to another. Work is a process in which such a change occurs by the action of a force moving, at least partly, in its own direction, and the term "work done" means the amount of energy changed in this process.

When the point of application of a constant force F moves through a distance s in the same direction as F, the work done is proportional to both F and s. Then, in suitable units,

$$W = Fs. \tag{4.1}$$

In any one system of units, all forms of energy may be measured in terms of the same basic unit, although for special purposes various units may be used. The basic unit of energy or work must agree with eqn. (4.1) when F and s are unit values. Thus unit energy is changed when unit work is done by the movement of the point of application of unit force through unit distance in its own direction.

In the M.K.S. system, $W = 1$ joule (J), when $F = 1$ newton (N) and $s = 1$ metre (m).

When an object is raised, energy must be supplied, as work is done against the weight of the object. If the object is lowered again, energy is released, as work is done *by* the weight. If the object is kept at rest at the upper level, e.g. by a shelf, the energy is stored in that situation. This form of stored energy is gravitational potential energy (E_p).

If an object is raised from rest at one level to rest at another there must be acceleration and deceleration. The vertical lifting force must therefore vary, being greater while the velocity is increasing than while it is decreasing. However, the average value of this vertical force F_{av} exactly equals the weight mg when rising from rest to rest. Then if h is the height through which the object is raised, eqn. (4.1) gives $W = F_{av}s = mgh$. This is the amount of energy stored in the form of potential energy.

$$\therefore\ E_p = mgh. \tag{4.2}$$

Solution

To find the height through which the load is raised, the resultant force is calculated:

$$\begin{aligned} F = \text{lifting force} - \text{weight} &= 1000-(60\times 9.81) \\ &= 1000-589 = 411\ \text{N}. \end{aligned}$$

When F is applied to the mass of 60 kg, the resulting acceleration is

$$a = F/m = 411/60 = 6.85\ \text{m s}^{-2}.$$

The distance moved in 0.40 s is given by

$$s = ut+\tfrac{1}{2}at^2 = 0+(\tfrac{1}{2}\times 6.85\times 0.40\times 0.40) = 0.548 \text{ m}.$$

$$\therefore\ E_p \text{ acquired} = mgh = 60\times 9.81\times 0.548 = 320 \text{ J}.$$

Note that the gain of E_p is not equal to the work done by the force of 1000 N, i.e. 1000×0.548 = 548 J. The difference of 228 J is due to the production of kinetic energy E_k as the load accelerates. Thus the total work done by the lifting force in this case equals the gain of E_p plus the gain of E_k.

4.2. Worked Example

A vehicle of mass 900 kg accelerates horizontally and uniformly from rest to a speed of 15 m s^{-1} in a time of 12 s, against a resistance of 200 N. Calculate (a) the work done against the resistance, (b) the kinetic energy acquired, and (c) the total work done by the engine.

Introduction

Kinetic energy is energy which is due to the motion of a body, and is acquired by a body in the process of acceleration. The energy comes from the source or sources contributing to the accelerating force.

Thus the gain of kinetic energy E_k equals the work done by the resultant accelerating force F moving through the distance s over which the acceleration occurs. But, from eqn. (3.2), $F = ma$.

$$\therefore\ \text{gain of } E_k = Fs = mas.$$

Now from eqn. (1.3), $as = \tfrac{1}{2}v^2-\tfrac{1}{2}u^2,$

$$\therefore\ \text{gain of } E_k = mas = \tfrac{1}{2}mv^2-\tfrac{1}{2}mu^2, \qquad (4.3)$$

where (mas) is the energy supplied to the mass, ($\tfrac{1}{2}mv^2$) is the final E_k, and ($\tfrac{1}{2}mu^2$) is the initial E_k. It may be shown that the gain of E_k is given by ($\tfrac{1}{2}mv^2-\tfrac{1}{2}mu^2$) even when the resultant force and acceleration are not constant, provided the mass is constant.

The units of energy given in ex. 4.1 apply equally to kinetic energy if m, u, and v are in the standard units for the particular system, e.g. u must be in metres per second and m in kilograms to give $\frac{1}{2}mu^2$ in joules.

Solution

(a) Work done against constant resistance = (force of resistance)×s, where the distance

$$s = (\text{average speed})\times\text{time} = (\tfrac{15}{2})\times 12 = 90 \text{ m}.$$

∴ work done against resistance

$$= 200\times 90 = 1.8\times 10^4 \text{ J}. \qquad \text{(a)}$$

(b) Using eqn. (4.3), gain of $E_k = \frac{1}{2}mv^2 - \frac{1}{2}mu^2$ where $u = 0$, $v = 15 \text{ m s}^{-1}$, and $m = 900$ kg.

$$\therefore \text{ gain of } E_k = \tfrac{1}{2}\times 900\times 15^2 = 1.0\times 10^5 \text{ J}. \qquad \text{(b)}$$

(c) The work done by the engine is of the two kinds, (a) and (b), i.e. the production of heat in overcoming the resistance and the production of kinetic energy.

$$\therefore \text{ total work} = (1.8\times 10^4)+(1.0\times 10^5) = 1.2\times 10^5 \text{ J}. \qquad \text{(c)}$$

Alternatively, the total work may be calculated from the total force exerted by the engine and the distance covered. Thus the accelerating force

$$= F = ma = m(v-u)/t = 900(15-0)/12 = 1130 \text{ N}.$$

$$\therefore \text{ total force} = (\text{accelerating force})+(\text{force opposing resistance})$$
$$= 1130+200.$$

$$\text{Then total work} = (\text{total force})\times\text{distance} = (1130+200)\times 90$$
$$= (1.0\times 10^5)+(1.8\times 10^4) \text{ (as before)}.$$

4.3. A 5.0 kg hammer is lifted and allowed to fall freely from a height of 2.0 m. Calculate the kinetic energy and hence the velocity after falling to the original level again. [98 J; 6.3 m s^{-1}.]

4.4. A mass of 4.0 kg falls from a height of 100 m. Calculate the potential energy at that height and the kinetic energy with which the mass reaches the ground. Assume the air resistance has an average value of 1.0 N. [3.9×10^3 J; 3.8×10^3 J.]

4.5. An electron has kinetic energy of 1.0×10^{-20} J. Calculate how far it travels against a constant force of deceleration if its kinetic energy is reduced to 16% in a time of 1.0 μs. Mass of electron = 9.1×10^{-31} kg. [10 cm.]

4.6. A lorry of mass 5000 kg is travelling at a speed of 72 km h^{-1} on the level. Calculate the kinetic energy and the stopping distance using the maximum friction available. The coefficient of friction is 0.50. [1.0×10^6 J; 41 m.]

4.7. A force of 1500 N accelerates a mass of 100 kg from rest for a time of 0.20 s. Calculate the momentum and kinetic energy gained. Also calculate the distance required to stop the mass with a force of 1000 N. [300 kg m s^{-1}; 450 J; 0.45 m.]

4.8. A locomotive of mass 2.0×10^4 kg, moving at a speed of 4.0 m s^{-1}, collides with a stationary truck of mass 5.0×10^3 kg. They couple automatically and move on together. Find their common velocity and the heat produced by the impact. [3.2 m s^{-1}; 3.2×10^4 J.]

Power

4.9. Worked Example

Calculate the output power of a machine working at a constant rate and exerting a constant force which raises a load of 30 tonne by 5.0 m in 30 s. Also calculate the power wasted if the efficiency is 40%.

Introduction

Power is the rate at which a transformation or transfer of energy occurs. Thus in a process of mechanical work, the power is the rate at which the work is done. In most cases an appreciable amount of energy is wasted in the process, so it is necessary to distinguish between the input power, the useful output power, and the power wasted.

When the point of application of a constant force F moves a distance s in the direction of the force, the work done $W = Fs$. If also the time taken is t and the velocity is v, then the power

$$P = W/t = (Fs)/t = Fv. \quad (4.4)$$

In short, power = rate of working = force×velocity. Note that if the velocity varies while the force is constant, as in constant acceleration, then the power varies also, but the average power is the product of the force and the average velocity.

In M.K.S. units the unit of power is the watt (W), equal to 1 joule per second.

Solution

Since the speed is constant, the force exerted by the machine must equal the weight of the load which equals $30 \times 1000 \times 9.81 = 2.94 \times 10^5$ N, while the velocity equals $\frac{5}{30} = 0.167$ m s^{-1}.

Then the useful output power $= 2.94 \times 10^5 \times 0.167 = 4.9 \times 10^4$ W.

The efficiency is given as 40%; i.e. the useful output power is 40% of the total input power, and the other 60% of the input is wasted.

$$\therefore \text{ power wasted} = (\tfrac{60}{40}) \times (\text{useful output power})$$
$$= (\tfrac{3}{2}) \times 4.9 \times 10^4 = 7.4 \times 10^4 \text{ W}.$$

4.10. Calculate the power of a fire pump which can throw 2000 l. of water per minute to a height of 50 m. [16 kW.]

4.11. A car of mass 750 kg accelerates from a speed of 36 km h^{-1} to 72 km h^{-1} in a time of 7.5 s on a level road and with negligible resistance to motion. Calculate (a) the increase of kinetic energy, and (b) the average useful output power. [(a) 1.1×10^5 J; (b) 15 kW.]

4.12. A locomotive with four coupled driving wheels has a mass of 60 tonne evenly distributed on them. Calculate the maximum tractive power possible at a speed of 10 km h^{-1} when the coefficient of friction between the wheels and the rails is 0.20. [330 kW.]

4.13. The turbines of a hydroelectric power plant have a supply of water of 10^6 kg min^{-1} falling from an effective height of 7.5 m. Calculate (a) the power input, and (b) the useful power output when the efficiency is 40%. [(a) 1.2 MW; (b) 0.49 MW.]

4.14. A lorry accelerates uniformly from rest to 20 km h^{-1} in a time of 3.0 s using an average power of 40 kW. Assuming negligible resistance, calculate (a) the accelerating force, (b) the work done, and (c) the peak power. [(a) 1.4×10^4 N; (b) 1.2×10^5 J; (c) 80 kW.]

Movement on Gradient

4.15. Worked Example

A mass of 0.50 kg is pulled up a gradient of 30° at a constant speed of 0.75 m s^{-1} against friction of 1.0 N. Calculate (a) the power required, and (b) the distance in which the mass would come to rest if the applied force were removed.

Introduction

No work is done by or against a force when its point of application moves in a direction perpendicular to the force. For example, no work is done by or against the weight of an object if it slides horizontally: there is no change of potential energy when there is no change of height.

If the point of application of the force F moves from A to B through a distance s at a constant angle to the direction of the force, the work done may be stated in terms of either a component of the force or a component of the displacement. Thus in Fig. 4.1

the force F is resolved into two components only one of which accounts for any work since the other has no movement in its own direction. Alternatively, in Fig. 4.2 the displacement s is resolved into two components, only one of which accounts for any work since the other is perpendicular to the force. The work done is the

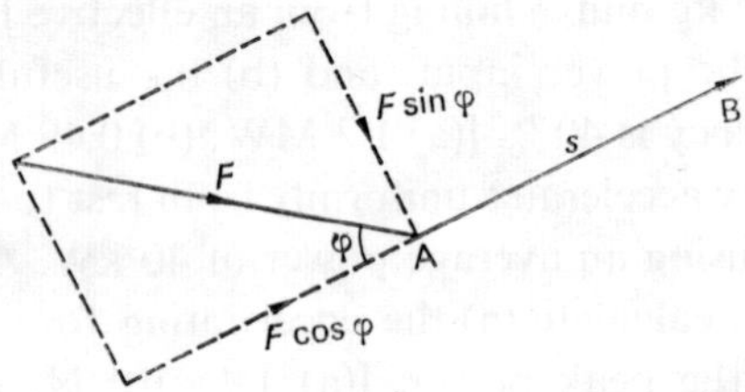

FIG. 4.1. Resolution of force in direction of displacement.

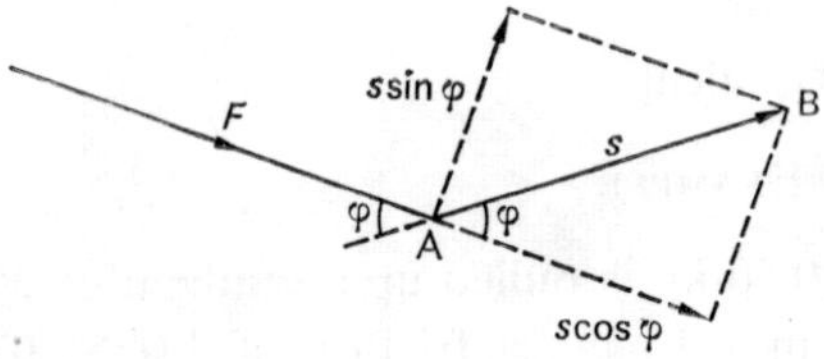

FIG. 4.2. Resolution of displacement in direction of force.

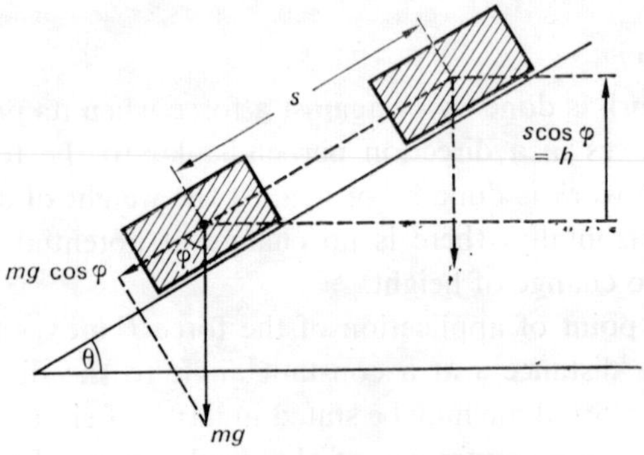

FIG. 4.3. Ex. 4.15.

same in both cases, i.e.

$$W = (F \cos \varphi)s = F(s \cos \varphi). \qquad (4.5)$$

In the particular question, when the mass m moves a distance s up the gradient, as in Fig. 4.3, the work done against the weight mg can be found in these two ways.

Thus $W = mg \cos \varphi s = mgs \cos \varphi$, and each expression equals the gain of $E_p = mgh$.

It may also be seen from Fig. 4.3 that $\sin \theta$ may be used in place of $\cos \varphi$ since $\theta + \varphi = 90°$.

Solution

(a) During an interval of 1 s, the gain of potential energy,

$$E_p = mgh = 0.50 \times 9.81 \times 0.75 \times \sin 30° = 1.84 \text{ J}.$$

In the same time, the work done against friction force F_f over the distance s in the direction opposite to the friction

$$= F_f s = 1.0 \times 0.75 = 0.75 \text{ J}.$$

$\therefore$ total energy used in 1 s $= 1.84 \times 0.75 = 2.59$ J.

$$\therefore \text{power} = 2.6 \text{ W} \quad (2 \text{ sig. fig.}). \qquad \text{(a)}$$

Alternatively, since the total driving force must balance both the friction and a component of the weight,

$$\text{driving force} = F_f + mg \sin \theta$$
$$= 1.0 + (0.50 \times 9.81 \times 0.50) = 3.45 \text{ N}.$$

$\therefore$ total work done per second $= 3.45 \times 0.75 = 2.59$ J (again).

(b) If the applied force were removed, the kinetic energy would immediately start to decrease as it became converted to potential energy and to heat due to friction.

Initial energy $E_k = \frac{1}{2}mv^2 = \frac{1}{2} \times 0.50 \times (0.75)^2 = 0.141$ J.

If s_2 = stopping distance, increase of E_p while stopping

$$= mgs_2 \sin\theta = 0.50\times 9.81\times s_2\times \sin 30° = 2.45s_2.$$

Also work done against friction while stopping $= F_f s_2 = 1.0\, s_2$.

Then, since energy cannot be created or destroyed, the initial energy equals the total final energy:

$$\therefore\ 0.141 = 2.45s_2+1.0s_2.$$

$$\therefore\ s_2 = 0.141/3.45 = 0.041 \text{ m} = 4.1 \text{ cm}. \qquad \text{(b)}$$

4.16. A man carries 10 l. of petrol in a 1 kg can at a speed of 6.0 km h^{-1} for a distance of 0.50 km up a hill of 1 in 10. Calculate the work done and the power used on the load. Assume relative density of petrol = 0.80. [4.4×10^3 J; 15 W.]

4.17. A vehicle of mass 1000 kg approaches a hill of 1 in 15 at a speed of 48 km h^{-1}. Calculate the kinetic energy at the foot of the hill. How far up the hill would the vehicle run with the engine off and with negligible resistance to motion? [8.9×10^4 J; 140 m.]

4.18. A car of mass 800 kg accelerates from rest to a speed of 24 km h^{-1} in a distance of 30 m up a slope of 10°. The frictional resistance to motion = 200 N. Calculate the work done by the engine. [6.5×10^4 J.]

4.19. A bulldozer pushes a block of masonry of mass 1000 kg at a constant speed of 2.0 m s^{-1} for a distance of 5.0 m up a slope of 30°. The frictional resistance to sliding is 5000 N. Calculate (a) the kinetic energy of the masonry, (b) the work done, and (c) the power supplied. [(a) 2000 J; (b) 5.0×10^4 J; (c) 20 kW.]

4.20. A mass of 50 kg slides from rest down an incline of vertical height 3.0 m and slope length 5.0 m. The kinetic energy at the base is 10 J. Calculate (a) the coefficient of friction, and (b) the minimum extra force needed to halve the time taken to reach the base. [(a) 0.24; (b) 600 N.]

CHAPTER 5

MACHINES

Lifting

5.1. Worked Example

A belt exerts a continuous effort of 100 N on the driving pulley of a hoist which raises a load of 50 kg at a speed of 0.60 m s^{-1}. If the belt transmits a power of 0.50 kW, calculate (a) the mechanical advantage, and (b) the efficiency of the hoist.

Introduction

A machine is a device which can do work in the particular way required when drawing on a supply of energy which would be unsuitable or inconvenient without the device. Thus in the present example the simple machine shown in Fig. 5.1 consists of a pulley, drum, and rope with the necessary shaft and bearings. It operates between the belt and the load so as to raise the load with the required force and at the required rate. The belt which transmits the energy from some source, shown only by another pulley, could not raise the load without some such device as the winding drum and rope.

The "mechanical advantage" is the ratio of the force exerted on the load to the effort exerted on the machine, or briefly,

$$\text{mechanical advantage} = \text{load/effort}, \tag{5.1}$$

where load and effort are expressed in the same units. Thus one of

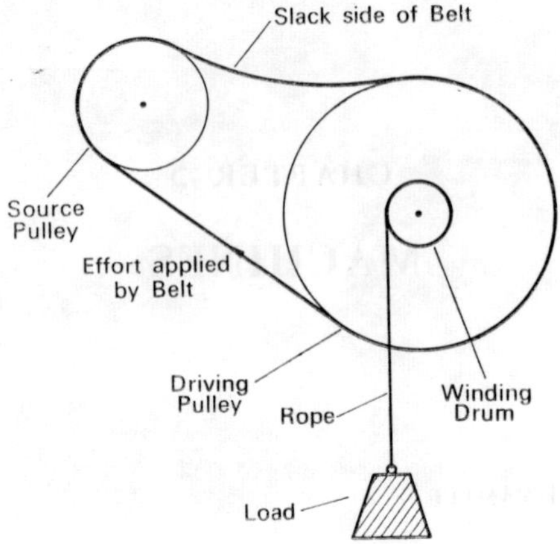

FIG. 5.1. Ex. 5.1.

the common requirements of a machine is to provide a high mechanical advantage so that a heavy load may be raised by a small effort.

Since all machines waste energy to some extent, the useful work done is less than the energy drawn from the supply. Then the efficiency,

$$\eta = \text{(useful output energy)/(input energy in same time)}$$
$$= \text{(useful output power)/(input power).} \qquad (5.2)$$

Solution

(a) To compare the load and the effort in the same units, the corresponding forces in newtons are considered. Thus, using (5.1),

$$\text{mechanical advantage} = (50\times 9.81)/100 = 4.9.$$

(b) Using eqn. (4.4),

$$\text{useful output power} = \text{(load force)}\times\text{(velocity of load)}$$
$$= (50\times 9.81)\times 0.60 = 294 \text{ W}.$$

Then the efficiency given by (5.2), with an input power of 500 W, is

$$294/500 = 0.59 = 59\%.$$

5.2. Worked Example

A wheel and axle are used to wind a rope which raises a load of 750 kg steadily by 1.0 m in 15 s. The handle on the wheel is at a radius of 50 cm, and the diameter of the axle on which the rope winds is 5.0 cm. The efficiency is 75%. Calculate (a) the velocity ratio, (b) the speed of the handle, (c) the mechanical advantage, (d) the work done by the effort, and (e) the turning moment of friction about the axle which would account for the stated efficiency.

Introduction

Consider the constant force of effort F_e to be applied tangentially to a circle of radius R as shown in Fig. 5.2, and the load to

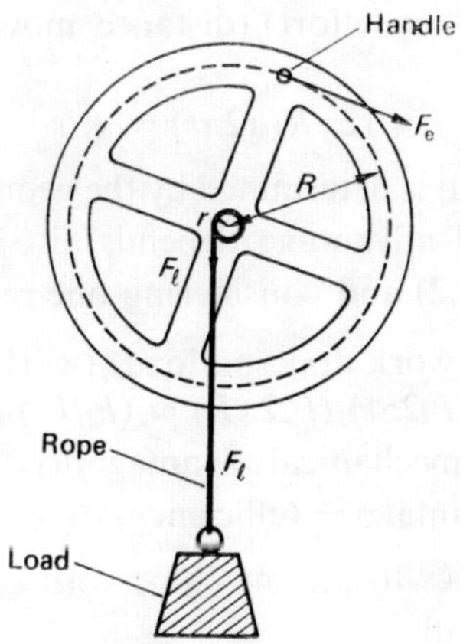

FIG. 5.2. Ex. 5.2.

hang on the rope which by its tension exerts the load force F_l upward on the load and tangentially downward on the axle at the smaller radius r. Then in one revolution the work done on the wheel by F_e is $F_e 2\pi R$, and the work done on the load is $F_l 2\pi r$.

Since the turning moments of these forces are F_eR and F_lr, the work done in one revolution in each case is

$$2\pi\times(\text{turning moment}). \tag{5.3}$$

Similarly, the work done in one revolution against the turning moment of friction is $2\pi\times$(turning moment of friction).

If there are no other turning moments but those due to effort, load, and friction, for constant speed,

$$(\text{moment of effort}) = (\text{moment of load}) + (\text{moment of friction}). \tag{5.4}$$

Then multiplying each bracket by 2π gives

$$(\text{work by effort}) = (\text{work on load}) + (\text{work against friction}), \tag{5.5}$$

agreeing with the law on conservation of energy.

Since the circumferences $2\pi R$ and $2\pi r$ are different, the velocities of effort and load are different. Thus there is a "velocity ratio"

= (velocity of effort)/(velocity of load)

= (distance moved by effort)/(distance moved by load in same time)

$$= (2\pi R)/(2\pi r) = R/r. \tag{5.6}$$

The velocity ratio is determined by the geometry of the machine, but the mechanical advantage depends also upon the efficiency. Thus, using eqn. (5.2) and considering one revolution of the axle,

$$\begin{aligned} \text{efficiency } \eta &= (\text{work done on load})/(\text{work done by effort}) \\ &= (F_l 2\pi r)/(F_e 2\pi R) = (F_l/F_e)/(R/r) \\ &= (\text{mechanical advantage})/(\text{velocity ratio}). \end{aligned}$$

$$\therefore \text{ mechanical advantage} = (\text{efficiency})\times(\text{velocity ratio}). \tag{5.7}$$

This equation applies to any machine with a single effort applied to move a single load.

Solution

(a) From eqn. (5.6), the velocity ratio $= 50/2.5 = 20$. (a)

(b) For a load velocity of $1.0/15$ m s^{-1},

$$\begin{aligned} \text{velocity of effort} &= (\text{velocity ratio})\times(1.0/15) \\ &= 20\times1.0/15 = 1.3 \text{ m s}^{-1} \quad (2 \text{ sig. fig.}). \end{aligned} \qquad \text{(b)}$$

(c) For an efficiency of 75% and a velocity ratio of 20, eqn. (5.7) gives

$$\text{mechanical advantage} = 0.75\times20 = 15. \qquad \text{(c)}$$

(d) The work done on the load in 15 s is found from

$$F_l = 750\times9.81 \text{ N and distance moved} = 1.0 \text{ m}.$$

$$\therefore \text{ work on load} = 750\times9.81\times1.0 \text{ J}.$$

Then, with $\eta = 0.75$,

$$\text{work done by effort} = (750\times9.81\times1.0)/0.75 = 9.8\times10^3 \text{ J}. \quad \text{(d)}$$

(e) From the load force and the mechanical advantage of 15 found in (c), the effort $= (750\times9.81)/15 = 490$.

Then, equating moments as eqn. (5.4) gives

$$(490\times0.50) = (750\times9.81\times2.5\times10^{-2})+(\text{moment of friction}).$$

$$\therefore \text{ moment of friction} = 245-184 = 61 \text{ N}. \qquad \text{(e)}$$

5.3. A machine can raise a load of 2000 kg through a height of 10 m in 0.25 min using an input power of 20 kW. Calculate the efficiency. [65%.]

5.4. The point of application of an effort of 100 N moves through a distance of 2.0 m in the direction of the effort when applied to a machine to raise a load of 1.0 tonne by 1.0 cm. Calculate the mechanical advantage, velocity ratio, and efficiency. [98, 200, 49%.]

5.5. A crane lifts a load of 0.50 tonne through a height of 5.0 m in a time of 3.0 s by winding in a cable at a rate of 5.0 m s^{-1} using an input power of 15 kW. Calculate the wastage of energy in that time, the mechanical advantage, and the efficiency. [2.0×10^4 J; 1.6; 55%.]

5.6. Calculate the effort required tangential to the rim of a wheel of 30 cm diameter, mounted on an axle of 2.0 cm diameter, to raise a load of 25 kg at the end of a thin rope wound around the axle. Assume (a) no friction, and (b) a total friction force of

50 N acting around the circumference of the axle. Also, in case (b), calculate the mechanical advantage, velocity, ratio, and efficiency. [(a) 16 N; (b) 20 N, 12, 15, 83%.]

5.7. A simple screw jack has a pitch of 0.50 cm and a handle of length 0.50 m. Calculate the force required on the handle to raise a load of 1.0×10^3 kg with efficiency of 25%. [62 N.]

Rotation and Power

5.8. Worked Example

The drum of a friction dynamometer has a diameter of 20 cm and is coupled to a motor shaft rotating at a constant rate of 2000 rev min^{-1}. The steady forces at the ends of the friction band are 500 N and 2500 N. Calculate (a) the output torque of the motor, and (b) the output power of the motor.

Introduction

The power transmitted by the rotation of a shaft depends upon its rate of rotation and upon the turning moment of the forces which it conveys. At any transverse cross-section of the shaft, the part on one side drives the other by exerting many forces between the particles of the material with a resultant turning moment called the "torque" in the shaft.

The output torque and output power of a motor with an output shaft can be measured by loading the shaft with a type of friction brake called a friction dynamometer. A drum of radius R is mounted on the output shaft, and a belt passing over this drum carries a load at one end and is hooked to a spring balance at the other (see Fig. 5.3).

When the shaft turns against the direction of the load force F_l, the spring balance indication drops to some approximately steady value F_s. Then, considering the belt to be in equilibrium, the fric-

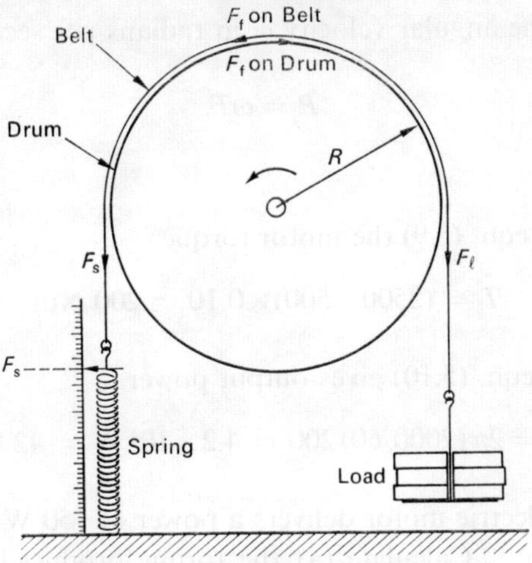

FIG. 5.3. Ex. 5.8.

tion force F_f between the belt and the moving drum just balances the other forces holding the belt.

$$F_f = F_l - F_s.$$

The work done against F_f in each revolution $= F_f 2\pi R$.

If the rate of rotation of the wheel is n in rev s^{-1}, the power or rate of doing work is given by

$$P = F_f 2\pi Rn = 2\pi Rn(F_l - F_s). \tag{5.8}$$

But when the drum speed is constant, the torque T in the motor shaft must equal the turning moment of the friction on the drum.

$$\therefore\ T = F_f R = (F_l - F_s)R. \tag{5.9}$$

Then eqn. (5.8) gives

$$P = 2\pi nT. \tag{5.10}$$

Note that n must be the rate of rotation in rev s^{-1}, not rev min^{-1}, if P is to be in watts and T in newton metres (Nm). Then

$2\pi n$ equals the angular velocity ω in radians per second (rad s^{-1}), and

$$P = \omega T. \tag{5.11}$$

Solution

(a) From eqn. (5.9) the motor torque

$$T = (2500-500)\times 0.10 = 200 \text{ Nm}. \tag{a}$$

(b) Then eqn. (5.10) gives output power

$$P = 2\pi(2000/60)\,200 = 4.2\times 10^4 \text{ W} = 42 \text{ kW}. \tag{b}$$

5.9. An electric motor delivers a power of 750 W at a speed of 3000 rev min^{-1}. Calculate (a) the torque produced, and (b) the diameter of the winding drum required to enable the motor to raise a load of 10 kg by a cord winding on to this drum at the stated speed of rotation. [(a) 2.4 Nm; (b) 4.9 cm.]

5.10. The torque produced by an electric motor at a speed of 2000 rev min^{-1} is 25 Nm. Calculate (a) the output power of the motor, and (b) the efficiency of the motor if the rate at which heat is developed in the motor is 1.0 kW. [(a) 5.2 kW; (b) 84%.]

Pulley Systems

5.11. Worked Example

Two pulley blocks, each containing two pulley wheels, are used as shown in Fig. 5.4 to raise a load. Different sizes of wheels, without flanges, are shown to make clear how the rope is wound. Calculate the load which may be raised steadily by an effort of 250 N when the efficiency is 70%. If the load speed is then 0.75 m s^{-1}, calculate the power wasted.

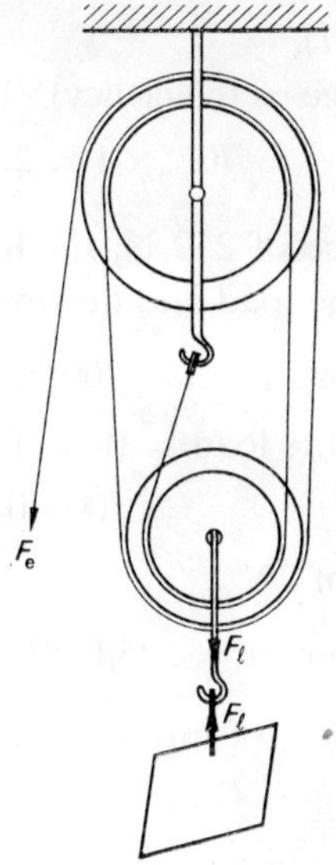

FIG. 5.4. EX. 5.11.

Introduction

The mechanical advantage cannot be calculated from the geometry of the system alone because it depends also upon the friction in the bearings of the pulley wheels ("sheaves") and the weight of the lower block. However, the velocity ratio depends upon the geometrical arrangement alone, and is readily calculated by considering the amount of "slack" produced if the lower pulley block were raised before hauling in the slack at the free end of the rope

Solution

In the particular case shown in Fig. 5.4, if the lower block were raised through a height h without hauling in the rope or turning the wheels, each of the four sections of rope supporting the lower block would slacken by the amount h. Then, to tighten the rope, a length $4h$ of slack would have to be hauled in. Hence the velocity ratio of the system is 4.

Then, using eqn. (5.7),

mechanical advantage = (efficiency)×(velocity ratio)

$= 70\% \times 4 = 2.8.$

$\therefore$ for an effort force of 250 N, the load force $F_l = 2.8 \times 250$ = 700 N, upward on the load and downward on the block.

$\therefore$ load mass = 700/9.81 = 71 kg.

The power used on the load = (load force)×(load speed)

$= 700 \times 0.75 = 525$ W.

Then for efficiency of 70%,

input power = 525/0.70 = 750 W.

$\therefore$ wasted power (= 30% of input) = 750 − 525 = 230 W (2 sig. fig.).

5.12. Worked Example

In a Weston differential pulley block the numbers of equally spaced recesses in the rims of the two wheels which are fixed together are 22 and 24 respectively. The mechanical advantage is 11.5 when raising a load of 200 kg including the mass of the lower block. Calculate (a) the efficiency, and (b) the ratio of effort forces required to raise and to lower the load respectively, assuming constant frictional resistance.

Introduction

In the Weston differential system, shown in Fig. 5.5, the upper block has two wheels of different radii r_A and r_B. An endless chain passes over both these wheels, in the rims of which there is a series of flats and recesses to match the links of the chain and so to prevent the chain slipping. The wheels are fixed together, so that if the force F_e, applied to the part D of the chain, turns these wheels through one revolution, part A rises by $2\pi r_A$ and part B falls by

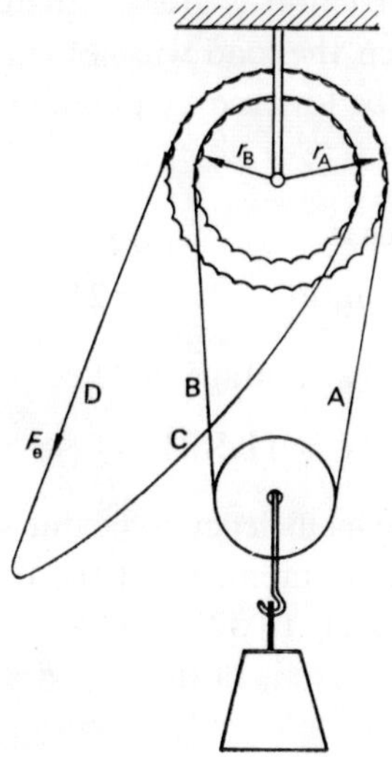

FIG. 5.5. Ex. 5.12.

$2\pi r_B$. Thus the total length of A and B hanging from the upper wheels is reduced by $2\pi r_A - 2\pi r_B$. The lower wheel turns so that this reduction is shared equally by A and B, and the load rises by $\pi r_A - \pi r_B$.

Then

$$\text{velocity ratio} = (\text{movement at } D)/(\text{movement of load})$$
$$= 2\pi r_A/(\pi r_A - \pi r_B) = 2r_A/(r_A - r_B). \qquad (5.12)$$

Since the circumferences of the upper wheels must be propor tional to the numbers of recesses N_A and N_B, so must the radii.

$$\therefore \text{ velocity ratio} = 2N_A/(N_A - N_B). \qquad (5.13)$$

Without friction the mechanical advantage would equal this velocity ratio, but it may be desirable to sacrifice some mechanical advantage by employing friction continuously, to prevent the system running back when released. Thus if the efficiency is less than 50% due solely to friction in the upper block, the maximum turning moment of friction is greater than the turning moment due to the load. Then the load will not run back when the chain is released, but may be lowered by pulling on the chain at C.

Solution

In the particular case, eqn. (5.13) gives

$$\text{velocity ratio} = 2\times 24/(24-22) = 24.$$

Then for a mechanical advantage of 11.5, eqn. (5.7) gives

$$\text{efficiency} = 11.5/24 = 0.48 = 48\%.$$

Now assuming the inefficiency to be due solely to friction in the upper block, the turning moments of the friction and of the load must be equal respectively to 52% and 48% of the moment due to the effort, relative to the axis of the upper wheels. Then, assuming the friction moment to be exactly reversed, the effort required at C to lower the load must give a turning moment equal to $(52-48)\%$ of the original effort required to raise the load.

But the efforts to raise and to lower the load act at the different radii r_A and r_B. Therefore the effort force to lower the load must be $(24/22)\times 4\%$ of the effort to raise it. Therefore these forces are in the ratio of 96 : 2200 = 12 : 275.

5.13. A pulley system has one wheel in the lower and two in the upper block. Calculate the effort required and the mechanical advantage when raising a load of 120 kg with an efficiency of 80%. [490 N; 2.4.]

5.14. An effort of 500 N, applied to a pulley system having a velocity ratio of 6, raises a load of 200 kg through a height of 5.0 m. Calculate (a) the work done by the effort, and (b) the energy wasted. [(a) 1.5×10^4 J; (b) 5.2×10^3 J.]

5.15. A man of mass 75 kg stands on a platform hanging from a pulley block, and raises himself and the platform by pulling on a rope. The rope passes round single-sheave pulley blocks, one supporting the platform and the other fixed overhead. Calculate the effort required assuming the mass of the platform to be (a) negligible, and (b) 30 kg. [(a) 250 N; (b) 340 N.]

5.16. A Weston differential pulley block has two wheels in the rims of which there are 24 and 25 recesses respectively. The larger wheel has an effective diameter of 25 cm. Calculate (a) the turning moment of friction on the upper wheels which would just prevent a load of 0.50 tonne from running back if the effort were relaxed, and (b) the effort required to raise the load with such friction. [(a) 12 Nm; (b) 200 N.]

Gears and Chains

5.17. Worked Example

A gear train consists of two pairs of gear wheels: a driver having 24 teeth engaging with a follower having 16 teeth, and, keyed to the same shaft as the follower, a second driver having 24 teeth engaging with a final follower having 12 teeth. For an output torque of 30 Nm at 250 rev min^{-1} and an efficiency of 25%, calculate (a) the input power, and (b) the input torque required.

Introduction

When the teeth of one gear wheel engage with those of another, the rate at which the teeth move must be the same for each wheel, as also for sprockets on wheels coupled by a chain. Then, ignoring any irregularities in the movements, the peripheral speeds v_A and v_B of the two wheels A and B are equal.

But, as in eqn. (1.9), $v = r\omega$, where r = radius and ω = angular velocity in radians per second.

$$\therefore \ \omega_A r_A = \omega_B r_B .$$

Then, since the radius of each wheel is proportional to the number N of its teeth,

$$\omega_A N_A = \omega_B N_B .$$

Therefore the "gear ratio"

$$\omega_A/\omega_B = N_B/N_A. \tag{5.14}$$

Thus the angular velocity ratio is the inverse of the ratio of the numbers of teeth on the wheels.

Solution

(a) As in eqn. (5.11),

$$\text{power} = \text{torque}\times\text{angular velocity}.$$

$$\therefore\ \text{output power} = 30\times(250\times 2\pi/60) = 250\pi \text{ W}.$$

Then for $\eta = 0.75$,

$$\text{input power} = 250\pi/0.75 = 1047 \text{ W} = 1.0 \text{ kW} \quad \text{(2 sig. fig.). (a)}$$

(b) Applying eqn. (5.14) to the two pairs of gear wheels shown in Fig. 5.6:

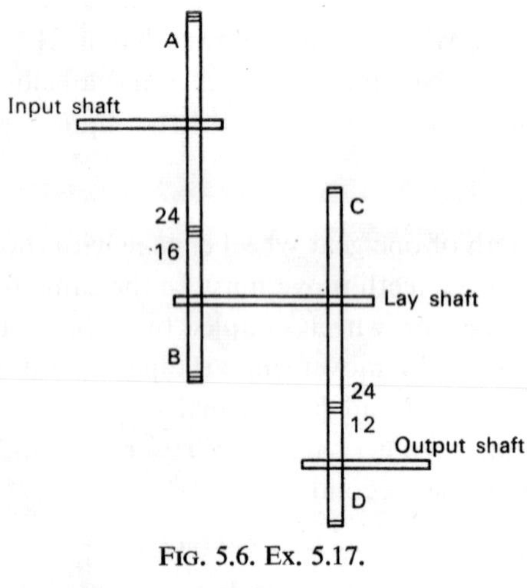

FIG. 5.6. Ex. 5.17.

for wheels A and B:

$$\omega_A/\omega_B = N_B/N_A = 16/24,$$

and similarly for wheels C and D:

$$\omega_C/\omega_D = N_D/N_C = 12/24.$$

But $\omega_B = \omega_C$ since both refer to the same shaft.

$$\therefore \left(\tfrac{24}{16}\right)\omega_A = \left(\tfrac{12}{24}\right)\omega_D = \left(\tfrac{12}{24}\right)(250\times 2\pi/60).$$
$$\therefore \omega_A = \left(\tfrac{16}{24}\right)\left(\tfrac{12}{24}\right)(500\pi/60) = 25\pi/9 \text{ rad s}^{-1}.$$

Then, as in eqn. (5.11),

torque = power/angular velocity = $1047/(25\pi/9) = 120$ Nm. (b)

5.18. A gearbox, having a speed reduction ratio of $3\frac{1}{2}:1$, delivers a power of 300 W at 1200 rev min^{-1} when the input torque is 0.80 Nm. Calculate the efficiency. [85%.]

5.19. A torque of 300 Nm is produced by a chain running at 3.0 m s^{-1} on a sprocket wheel of diameter 20 cm. Calculate the power transmitted and the work done in 350 rev. [9.0 kW; 6.6×10^5 J.]

5.20. A sprocket wheel of diameter 10 cm is driven by a chain from a wheel of diameter 4.0 cm, each wheel being on a shaft of diameter 1.0 cm. The tension in the driving side of the chain is 30 N, and that in the slack side is negligible. Taking the load and effort forces to act around the circumferences of the shafts, calculate (a) the velocity ratio, and (b) the power transmitted when the driving wheel turns at a rate of 100 rev min^{-1}, assuming 100% efficiency. [(a) 2.5; (b) 6.3 W.]

5.21. A bicycle has pedal cranks of length 18 cm, and wheels of diameter 65 cm. The chain wheel has 45 teeth and the rear sprocket wheel has 16. The effort on the pedal is applied perpendicular to the crank. Calculate the velocity ratio and the efficiency when the mechanical advantage is $\frac{1}{6}$. [0.20; 85%.]

CHAPTER 6

ELASTICITY

Helical Springs

6.1. Worked Example

Three identical helical springs hang vertically and share equally in supporting a load. If the load is increased by 6.0 kg the extension increases by 3.0 cm. Calculate the further extension of two springs if the third is removed and the remaining two share the load equally.

Introduction

Springs and material in many other forms are said to exhibit elasticity, referring to the ability of the material to return to its original size and shape when distorting forces are removed. This applies only up to a "limit of elasticity", beyond which the material would retain some distortion. In many cases it is found that the distortion produced by an applied force is proportional to the force, e.g. the extension of a simple spring balance is proportional to the load which hangs on it, but only up to a "limit of proportionality".

Although these limits refer to different effects, i.e. elasticity is not the same as proportionality, it is found that the limits are usually very close. This is implied in Hooke's law that extension is proportional to load up to the elastic limit.

Solution

When there are three springs, each is extended by 3.0 cm due to its one-third share of the load increase, i.e. due to $6.0/3 = 2.0$ kg. Then, assuming that the elastic limit is not exceeded and that Hooke's law applies, the proportion may be stated as $3.0/2.0 = 1.5$ cm kg^{-1}.

When only two springs support the load the share of each is 3.0 kg, and therefore the extension is $3.0 \times 1.5 = 4.5$ cm. Thus the extension is increased by $4.5 - 3.0 = 1.5$ cm.

6.2. Worked Example

The scale of an inaccurate spring balance is accurately divided in millimetres, but 1.0 mm inaccurately indicates a load of 10 g. For zero load the indication is 15 g, and for a true load of 500 g the indication is 490 g. Calculate (a) the stiffness of the spring, and (b) the true load force which gives an indication of 860 g. Assume normal acceleration of gravity throughout.

Introduction

A simple spring balance is usually calibrated in units of mass because its main purpose is to determine the masses of loads hung on it. However, the extension of the spring depends primarily upon the force exerted on it by the load rather than upon the mass of the load. This force varies not only with the mass but with the acceleration of gravity or other acceleration (see ex. 3.2). Thus calibration in absolute units of force would be more correct, but is not common because of the consequent calculation required to determine the mass.

In the present example it is assumed that only two errors are involved: (1) a zero error due to incorrect adjustment, and (2) a stiffness error due to incorrect choice of spring to suit the scale or vice versa. The stiffness is defined as the load force per unit extension of the spring within the range of Hooke's law.

Solution

The zero error shown in Fig. 6.1 is assumed to be due to the upper end of the spring being set 1.5 mm too low relative to the scale.

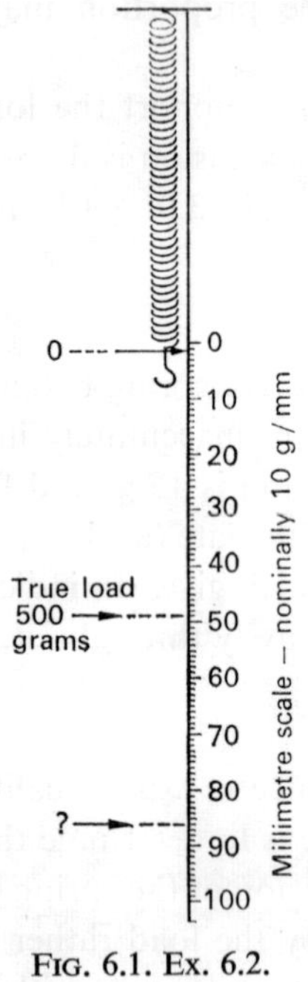

FIG. 6.1. Ex. 6.2.

The stiffness is given by the extension from 1.5 mm to 49 mm due to a load force of 500 gf. Then, in absolute units,

$$\text{stiffness} = 0.500\times 9.81/(49-1.5) = 0.103 \text{ N mm}^{-1}. \quad \text{(a)}$$

The indication of 860 g occurs when the extension $= 86.0-1.5 = 84.5$ mm. Then the true load force

$$= 84.5\times 0.103 = 8.7 \text{ N}. \quad \text{(b)}$$

6.3. A spring hanging vertically has a length of 7.25 cm when loaded with a mass of 50 g and 8.50 cm with 100 g. Calculate (a) the stiffness, and (b) the load which stretches it to 10.0 cm. Assume $g = 9.8 \text{ m s}^{-2}$. [(a) 39 N m^{-1}; (b) 160 g.]

6.4. A spring balance indicates a mass of 10.0 g on zero load and 60.3 g with a true load of 50.0 g. Calculate the true load and load force when the indication is 100.0 g. Assume $g = 9.81 \text{ m s}^{-2}$. [89.5 g; 0.877 N.]

Young's Modulus

6.5. Worked Example

A cable consists of 19 steel strands, each of diameter 1.5 mm and length 25 m. Calculate the tension which would cause an extension of 0.50 cm, assuming the extension to be due entirely to stretching of the strands, and assuming Young's modulus for the material to be $2.2 \times 10^{11} \text{ N m}^{-2}$.

Introduction

If a wire of material such as steel is stretched, and a graph is plotted showing the extension compared with the tension, a straight line is obtained within the elastic limit in accordance with Hooke's law. A different length or gauge of wire or a different material would give a different slope to the graph (see Fig. 6.2). Different lengths stretch in proportion to their lengths; different gauges stretch in inverse proportion to their cross-section areas.

To allow conveniently for the effects of different sizes, tensile stress and tensile strain are defined:

$$\text{tensile stress} = \text{tension force/cross-section area} = F/A; \qquad (6.1)$$

$$\text{tensile strain} = \text{extension/original length} = \Delta l/l_0. \qquad (6.2)$$

Note that stress has units like pressure but strain is a pure ratio.

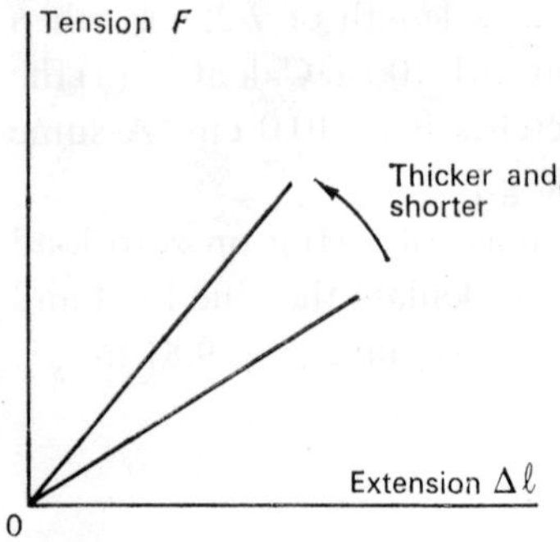

FIG. 6.2. Tension–extension curves.

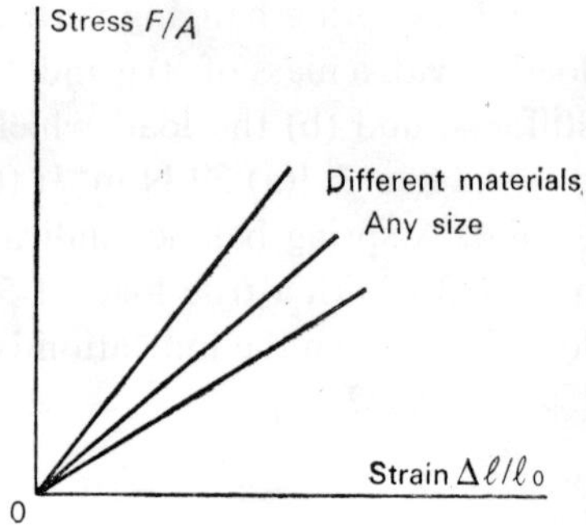

FIG. 6.3. Stress–strain curves.

Plotting stress against strain, as in Fig. 6.3, gives only one graph for each material whatever the size of the wire or other test piece. The gradient of the stress–strain graph along the straight part shown, up to the limit of proportionality, is called Young's modulus,

$$E = \text{tensile stress}/\text{tensile strain} = (F/A)/(\Delta l/l_0). \qquad (6.3)$$

Solution

Following the definitions of (6.1) and (6.2):

$$\text{tensile strain} = 0.50\times 10^{-2}/25 = 2.0\times 10^{-4}$$

and $$\text{tensile stress} = F/\pi(0.75\times 10^{-3})^2 \text{ N m}^{-2},$$

where F is the tension in each strand.

Then, from the definition of the modulus, eqn. (6.3),

$$\text{stress} = E\times\text{strain} = 2.2\times 10^{11}\times 2.0\times 10^{-4} = 4.4\times 10^{7} \text{ N m}^{-2}.$$

Comparing the two expressions for stress gives

$$F = 4.4\times 10^{7}\times\pi(0.75\times 10^{-3})^2 = 78 \text{ N}.$$

So for 19 strands, total tension $= 19\times 78 = 1500$ N.

6.6. A wire of length 4.0 m and diameter 1.0 mm extends by 6.0 mm when supporting vertically a load of 25 kg. Calculate the value of Young's modulus for the material. [2.1×10^{11} N m^{-2}.]

6.7. A 10 m length of steel wire is required to extend by no more than 1.0 mm when carrying a load of 1.0 kg in direct tension. Calculate the minimum diameter required. Ref.: steel. [0.75 mm.]

6.8. A wire of length 3.0 m and diameter 0.50 mm extends by 1.0 mm under a certain tensile load. Calculate the extension of a 5.0 m length of 0.30 mm diameter wire of the same material carrying four-fifths of the load. [3.7 mm.]

6.9. Part of the framework of a building is an H-section girder of length 5.0 m. The cross-section has overall dimensions of 15×15 cm, with an average thickness of 2.5 cm. A load of 75 tonne is supported in compression. Calculate the reduction of length, assuming Young's modulus to apply in compression. Ref.: steel. [1.7 mm.]

6.10. A steel wire of length 1.0 m and diameter 0.50 mm is just taut between rigid posts. Calculate how far the centre should be pulled sideways to produce a tension of 1.0 N in the wire. Ref.: steel. [3.4 mm.]

Compound Bars

6.11. Worked Example

A load of 5000 kg hangs at the end of a compound bar 50 cm long with axis vertical. The bar has a steel core of 2.0 cm diameter in a closely fitting brass tube of equal length and of 3.0 cm outside diameter. Calculate the extension of the bar and the load force carried by each metal. Ref.: brass and steel.

Solution

If the bar stretches without relative slipping of the brass and steel, the extension and strain are the same for each material.

Now tensile stress $\sigma = E\Delta l/l_0$, where $E =$ Young's modulus and $\Delta l/l_0 =$ tensile strain.

$\therefore$ for brass $\sigma_b = 9.0\times10^{10}\Delta l/0.50 = 18\times10^{10}\Delta l$ N m^{-2},

and for steel $\sigma_s = 22\times10^{10}\Delta l/0.50 = 44\times10^{10}\Delta l$ N m^{-2}.

But tension F = stress$\times$cross-section area.

$$\therefore \text{ for the brass } F_b = 18\times10^{10}\Delta l\times\pi(1.5^2-1.0^2)\times10^{-4} = 7.08\times10^{7}\Delta l \text{ N},$$

and
$$\text{for the steel } F_s = 44\times10^{10}\Delta l\times\pi 1.0^2\times10^{-4} = 13.8\times10^{7}\Delta l \text{ N}.$$

$$\text{Total tension} = F_b+F_s.$$

$$\therefore 5000\times9.81 = 20.9\times10^{7}\Delta l. \quad \therefore \Delta l = 2.34\times10^{-4}\text{ m}.$$

Then
$$F_b = 7.08\times10^{7}\times2.34\times10^{-4} = 1.7\times10^{4}\text{ N}$$

and
$$F_s = 13.8\times10^{7}\times2.34\times10^{-4} = 3.2\times10^{4}\text{ N}.$$

6.12. A compound bar consists of a brass tube fitting closely on a steel core of diameter 2.0 cm. Calculate the maximum thickness of the tube if it is to carry no more than half the load force carried by the core when the bar is in tension. Ref.: brass and steel. [0.49 mm.]

6.13. A steel wire is joined end to end to a brass wire of half the length and double the diameter. If the diameter of the steel were doubled, what effect would this have on the total extension of the wires when loaded in tension? Ref.: brass and steel. [58% decrease.]

Bulk Modulus

6.14. Worked Example

Given that the bulk modulus of mercury is 2.5×10^{10} N m^{-2}, calculate (a) the compressibility of mercury, and (b) the pressure required to reduce the volume of mercury by one part in a million.

Introduction

If a specimen of solid or liquid is subjected to a uniform pressure over its entire surface area, the resulting reduction of volume $-\delta V$ depends upon the original volume V_0, the pressure p, and the nature of the material. The relationship follows Hooke's law in the region of elasticity, and is expressed by the bulk modulus

$$K = \text{volumetric stress/volumetric strain}$$

where volumetric stress $= p$, and volumetric strain $= -\Delta V/V_0$. Thus

$$K = p/(-\Delta V/V_0).$$

Alternatively the reciprocal $1/K$ is called the compressibility.

This modulus for a solid differs from Young's modulus for the same material under only a longitudinal compression because, when the pressure extends over the entire surface, the transverse expansion which accompanies longitudinal compression is prevented and a smaller compression occurs in all directions.

Solution

(a) Compressibility $=$ volumetric strain/pressure

$$= 1/K = 1/2.5\times10^{10} = 4.0\times10^{-11}\ \text{m}^2\ \text{N}^{-1}. \quad \text{(a)}$$

(b) Pressure required $= K\times$volumetric strain

$$= 2.5\times10^{10}\times10^{-6} = 2.5\times10^{4}\ \text{N m}^{-2}. \quad \text{(b)}$$

6.15. Calculate the pressure head of water required to reduce the volume of water by 2 parts in a 1000. Assume the bulk modulus of water to be 2.2×10^9 N m^{-2}. [450 m.]

6.16. Assuming Boyle's law for air, calculate (a) the pressure required at constant temperature to make a $\frac{1}{2}$% reduction of volume at s.t.p., and hence (b) the isothermal bulk modulus of air at s.t.p. [(a) 500 N m^{-2}; (b) 1.0×10^5 N m^{-2}.]

Torsion

6.17. Worked Example

In a static torsion test a nickel wire of length 25 cm and diameter 1.0 mm hangs vertically with a drum of diameter 5.0 cm attached coaxially at the lower end. A torque is produced in the wire by a couple applied by two cords wound around the drum and leaving tangentially from opposite sides. The drum is rotated through 55° when the tension in each cord is 60 gf. Calculate the modulus of rigidity or shear modulus.

Introduction

When one end of the wire is twisted relative to the other by the couple, torsional strain is produced because each longitudinal element of the wire undergoes shear strain. Figure 6.4 shows an

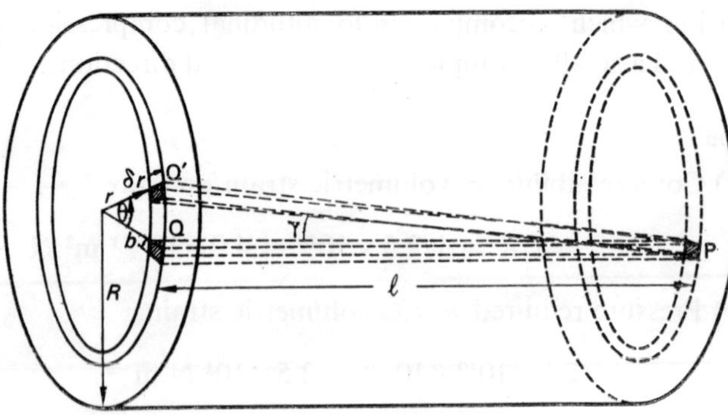

Fig. 6.4. Ex. 6.17.

element PQ' of the wire in this state of strain after one end Q' of the element has been moved along a small arc from the unstrained position Q. The shear strain tends to disappear elastically when the couple is removed, but is not a simple case of longitudinal or volumetric strain although all three types of strain are interrelated.

The shear strain of the element is the angle of distortion in radians, shown by γ in Fig. 6.4, where $\gamma = r\theta/l$ for small values of the angle of twist θ. The shear stress on this element $= F/b\delta r$, where F is the force which drives Q to Q', and $b\delta r$ is the cross-section area of the element.

Then the ratio (shear stress)/(shear strain), called either the modulus of rigidity or the shear modulus, $G = Fl/(r\theta b\delta r)$.

$$\therefore\ F = G(r\theta b\delta r)/l.$$

For the complete cylindrical element of radius r and thickness δr in Fig. 6.4, b becomes $2\pi r$, and the total turning moment of all forces like F around this circumference $= r\,G(r\theta 2\pi r\delta r)/l$.

Therefore for the complete wire of radius R, the torque

$$T = \int_0^R (2\pi G\theta/l)r^3 dr = (2\pi G\theta/l)R^4/4 = \pi G\,R^4\theta/2l. \qquad (6.4)$$

Solution

From the data,

$$R = 5.0\times 10^{-4}\text{ m},\quad l = 25\times 10^{-2}\text{ m},$$
$$\theta = 55\times\pi/180 = 0.96\text{ rad},$$

and $T = (60\times 9.81\times 10^{-3})\times(5.0\times 10^{-2}) = 2.94\times 10^{-2}$ N m.

Then, from eqn. (6.4),

$$G = 2lT/(\pi R^4\theta)$$
$$= 2\times 25\times 10^{-2}\times 2.94\times 10^{-2}/(\pi\times 625\times 10^{-16}\times 0.96)$$
$$= 7.8\times 10^{10}\text{ N m}^{-2}.$$

6.18. A steel shaft of diameter 1.0 cm and length 25 cm transmits a torque of 5.5 Nm. Assuming a shear modulus of 8.0×10^{10} N m^{-2}, calculate the angle of twist in the shaft. [1°.]

6.19. A certain galvanometer has a moving coil suspended by a quartz fibre. It is proposed to replace the fibre by a phosphor bronze wire which shall have the same torsional rigidity. Calculate the change in the factor of safety which relates the tensile strength to the tension supporting the weight of the coil. Assume the following data for the two materials:

Shear modulus (N/m^2): quartz, 3.0×10^{10}; phosphor bronze, 4.4×10^{10}.

Tensile strength (N/m^2): quartz, 2.5×10^{8}; phosphor bronze, 9.0×10^{8}. [$\times$3.]

Stored Energy

6.20. Worked Example

When a load of 10 kg hangs from the end of a vertical wire, the extension of the wire is 0.80 mm, and when one end is twisted relative to the other through one complete revolution the torque produced is 2.0×10^{-2} Nm. Calculate the energy stored in the wire in each case.

Solution

During the process of extension, the tension in the wire increases from zero to 10×9.81 N in proportion to the extension at each stage, and with a mean value of 5×9.81 N. Then the energy stored on account of the work done against this tension = (mean tension)$\times$(total extension)

$$= 5\times9.81\times0.80\times10^{-3} = 3.9\times10^{-2}\ \text{J}.$$

During the process of twisting, the torque increases in proportion to the angle of twist, with a mean value of 1.0×10^{-2} Nm. But when a torque or couple of turning moment T causes a rotation of θ radians, the work done is $T\theta$, since for each component force F having moment Fr the point of application moves a distance $r\theta$, and the work done is the product $Fr\theta$. Thus the energy stored on account of the work done in twisting the wire

$$= 1.0\times10^{-2}\times2\pi = 6.3\times10^{-2} \text{ J.}$$

6.21. A helical spring having an unstressed length of 10 cm extends to 15 cm when supporting a load of 50 g. Calculate the energy required to extend the spring from its natural length to a length of 18 cm. [0.031 J.]

6.22. Calculate the work done in producing an extension of 0.50 mm in a steel wire of length 3.0 m and diameter 1.0 mm. Ref.: steel. [7.2×10^{-3} J.]

6.23. Calculate the energy stored in the wire of ex. 6.22 when one end is twisted through two complete revolutions relative to the other, if the modulus of rigidity is 8.0×10^{10} N m^{-2}. [0.21 J.]

6.24. A spiral spring, standing vertically and having its lower end fixed, is shortened by 2.0 cm when supporting a load at rest upon it. The load is then pressed down by a further 4.0 cm and released. Calculate the greatest height to which the load rises from its position when released. [8.0 cm.]

CHAPTER 7

EXPANSION AND TEMPERATURE

Linear Temperature Scales

7.1. Worked Example

A mercury-in-glass thermometer is placed alongside a scale divided in millimetres. At 0°C the mercury level is at the point 13 mm and at 100°C it is at 178 mm. Calculate the temperature when the level is at 46 mm.

Introduction

If two substances which do not react chemically are brought into contact, the various collisions of their molecules may cause a resultant transfer of thermal kinetic energy, i.e. a flow of heat from one substance to the other. The substance which loses the heat is said to start at a higher temperature than that which gains it. To avoid the difficulties of measuring temperature directly in terms of thermal motion, various easily measurable properties of substances, such as volume and electrical resistance, are used. These properties vary suitably with temperature and so may be used to indicate temperature.

A thermometer is a device in which a convenient property indicates the temperature of the thermometer itself and therefore the temperature of any substance in thermal equilibrium with it. In the most common type of thermometer, the different expansions of mercury and glass with change of temperature result in the mercury thread moving along a scale marked on the glass.

Figure 7.1 displays the meaning of a linear scale of temperature for a mercury-in-glass thermometer. Here x is the distance of the mercury "level" from some convenient zero level. The straight line shows that the temperature θ is taken to be linearly related to x.

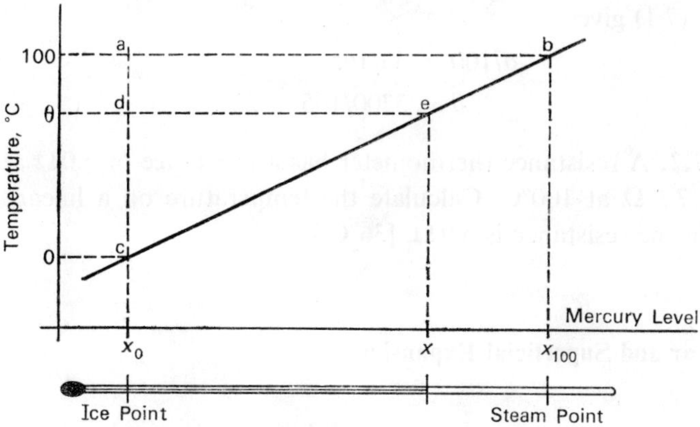

FIG. 7.1. Linear temperature scale.

θ can then be calculated using the ice and steam points which define the temperatures of 0° and 100° on the Celsius scale. Thus, from the similar triangles of Fig. 7.1, $dc/ac = de/ab$.

$$\therefore\ (\theta-0)/(100-0) = (x-x_0)/(x_{100}-x_0).$$

$$\therefore\ \theta = 100(x-x_0)/(x_{100}-x_0). \qquad (7.1)$$

Similarly, linear temperature scales can be devised using other materials or properties such as electrical resistance R or thermo-electric e.m.f. E in place of x, provided that the chosen property increases or decreases as heat is supplied, and reverses when and only when heat is removed. However, if all thermometers have linear scales, the indications are different on thermometers using different substances or properties which vary out of proportion with one another. When such differences are important, the ideal gas thermometer is taken as the standard, but in many cases the differences are negligible, as in the following problems.

Solution

When $x_{100}-x_0 = 178-13 = 165$ mm

and $x-x_0 = 46-13 = 33$ mm

eqn. (7.1) gives

$$\theta/100 = 33/165.$$

$$\therefore \theta = 3300/165 = 20°\text{C}.$$

7.2. A resistance thermometer has a resistance of 5.0 Ω at 0°C and 7.5 Ω at 100°C. Calculate the temperature on a linear scale when the resistance is 5.9 Ω. [36°C.]

Linear and Superficial Expansion

7.3. Worked Example

Calculate the lengths of brass rods *A* and *B* at a temperature of 300°C, if the length of *A* is 18.000 cm at 0°C, and that of *B* is 18.000 cm at 100°C. Ref.: coefficient of expansion.

Introduction

When a solid body is heated, its length and other linear dimensions increase approximately in proportion to the temperature rise as determined by a mercury-in-glass thermometer with a linear scale (see ex. 7.1).

Thus, assuming exact proportion,

$$\Delta l = \alpha l \Delta\theta, \tag{7.2}$$

where Δl = increase of length, l = length at the original temperature, $\Delta\theta$ = rise of temperature, and α = coefficient of linear expansion, depending on the particular material and the scale of temperature.

It is usually assumed that α is not dependent upon the original temperature, but this cannot be exactly true if the expansion is proportional to the original length, which depends upon the original temperature. However, the error is negligible when $\alpha\Delta\theta \ll 1$, and in practice a greater error would arise from a departure from the simple proportionality.

Now if θ_1 is the original temperature at which the length is l_1, and θ_2 is another temperature at which the length is l_2, then

$$\Delta l = l_2 - l_1, \quad \text{and} \quad \Delta\theta = \theta_2 - \theta_1.$$

Therefore from eqn. (7.2),

$$l_2 - l_1 = \alpha l_1(\theta_2 - \theta_1)$$

or

$$l_2 = l_1[1 + \alpha(\theta_2 - \theta_1)]. \tag{7.3}$$

Equation (7.3) may be used instead of calculating Δl separately from eqn. (7.2) and then adding l_1 to obtain l_2. But when Δl is very small, high numerical accuracy is obtained more easily by taking the two steps separately.

Solution

(a) Using eqn. (7.2) for rod A, with $\theta_1 = 0°C$, and $\alpha = 19\times10^{-6}$ deg C^{-1} from the reference tables,

$$\Delta l = 19\times10^{-6}\times18.000\times300 = 0.103.$$

Then

$$l_{300} = l_0 + \Delta l = 18.000 + 0.103 = 18.103 \text{ cm.} \tag{a}$$

(b) Using the same value of α again, but with $\Delta\theta = 300 - 100 = 200°C$, eqn. (7.2) gives

$$\Delta l = 19\times10^{-6}\times18.000\times200 = 0.0684.$$

Then for rod B,

$$l_{300} = 18.000 + 0.0684 = 18.068 \text{ cm.} \tag{b}$$

As a check on the error in using the same value of α when $\theta_1 = 100°C$ as when $\theta_1 = 0°C$, l_{300} for rod B may be recalculated using $\theta_1 = 0°C$.

Thus from eqn. (7.3), $l_{100} = l_0[1+(\alpha 100)]$.

$\therefore\ l_0 = l_{100}/[1+(100\alpha)] = 18.000/[1+(19\times 10^{-4})] = 17.966$ cm.

Then, for a rise to 300°C, (1) gives

$$\Delta l = 19\times 10^{-6}\times 17.966\times 300 = 0.102.$$

$$\therefore\ l_{300} = 17.966+0.102 = 18.068 \text{ cm}.$$

Thus the two results agree in the first 5 significant figures.

7.4. Worked Example

A steel plate at a temperature of 0°C has a hole of diameter 2.0 cm through which a rod is to pass. Calculate the necessary coefficient of linear expansion of the material of the rod to ensure that there is, at all temperatures, (a) a constant radial clearance of 1.0 mm between the rod and plate, and (b) a constant annular area of 0.50 cm² between rod and plate. Ref.: steel.

Introduction

When the diameter of a flat circular plate or the side of a flat square plate increases, so does the corresponding area.

Thus for the diameter or side, $l_2 = l_1(1+\alpha\Delta\theta)$.

Then for the area of the circle, $(\pi/4)l_2^2 = (\pi/4)\,l_1^2(1+\alpha\Delta\theta)^2$.

$$\therefore\ l_2^2 = l_1^2(1+2\alpha\Delta\theta+\alpha^2\Delta\theta^2) \tag{7.4}$$

for the circle, as also for the square.

The last term may be neglected since $\alpha^2\Delta\theta^2 \ll 1$. In the case of the square plate, this is apparent from Fig. 7.2, where the expansion of area shown along two sides of the square is $2l_1(l_1\alpha\Delta\theta)$, and the small square in the top corner has the negligible area $(l_1\alpha\Delta\theta)^2$.

Then, writing A for area and β for the coefficient of superficial (or area) expansion in place of 2α, eqn. (7.4) becomes

$$A_2 = A_1(1+2\alpha\Delta\theta) = A_1(1+\beta\Delta\theta). \tag{7.5}$$

Also $$\Delta A = \beta A\Delta\theta. \tag{7.6}$$

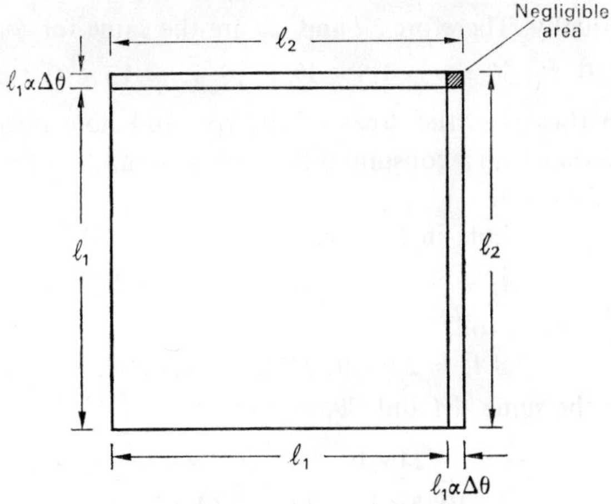

FIG. 7.2. Expansion of a flat square plate.

Note that eqns. (7.5) and (7.6) are of the same form as those for linear expansion in ex. 7.3.

Solution

(a) The diameter of the hole in the steel expands just as if it were filled with steel. Thus the diameter has a linear coefficient of expansion as given in the reference table for steel:

$$\alpha_h = 12\times 10^{-6} \text{ deg C}^{-1},$$

and $l_h = 2.0$ cm, where the suffix h refers to the hole.

Then, using eqn. (7.2), the increase of hole diameter is

$$\Delta l_h = \alpha_h l_h \Delta\theta = (12\times 10^{-6})\times 2.0\times\Delta\theta = 24\times 10^{-6}\times\Delta\theta.$$

To give the required clearance for part (a), the diameter of the hole at 0°C must be 2.0 cm − 2 mm = 1.8 cm.

$\therefore$ the increase of rod diameter $= \Delta l_r = \alpha_r l_r \Delta\theta = \alpha_r\times 1.8\times\Delta\theta$.

Now for (a) the diameters of the rod and hole are required to expand equally. Therefore Δl and $\Delta\theta$ are the same for each.

$$\therefore\ 24\times10^{-6}\times\Delta\theta = \alpha_r\times1.8\times\Delta\theta. \quad \therefore\ \alpha_r = 13\times10^{-6}\ \text{deg C}^{-1}.$$

(b) In this case, the areas of the rod and hole must expand equally to maintain a constant difference of area. $\therefore$ ΔA and $\Delta\theta$ are the same for each.

But, for the hole in the steel, $\beta_h = 2\alpha_h = 24\times10^{-6}$ deg C^{-1}.

$$\therefore\ \Delta A_h = \beta_h A_h \Delta\theta = 24\times10^{-6}(\pi/4)2.0^2\Delta\theta.$$

Similarly for the rod,

$$\Delta A_r = 2\times\alpha_r[(\pi/4)2.0^2-0.50]\Delta\theta.$$

Then, for the same ΔA and $\Delta\theta$,

$$24\times10^{-6}\times\pi = 2\times\alpha_r[\pi-0.50].$$

$$\therefore\ \alpha_r = 12\times10^{-6}\times3.14/2.64 = 14\times10^{-6}\ \text{deg C}^{-1}.$$

7.5. A metal rod has a length of 15.000 cm at 0°C, and 15.040 cm at 100°C. Calculate the lengths at (a) 150°C, and (b) 180°C. [(a) 15.060 cm; (b) 15.072 cm.]

7.6. A steel tyre is to be shrunk on to a wheel centre of diameter 90.00 cm at a temperature of 15°C. The tyre is bored to a diameter of 89.95 cm at 15°C. Calculate the temperature to which the tyre must be heated to allow an average clearance of 0.10 cm around the wheel centre as the tyre is placed in position. Ref.: steel. [250°C.]

7.7. A thermostat operates by the differential expansion of an invar rod in a brass tube. Both rod and tube have a length of 20 cm at 0°C. Calculate the differential expansion per unit rise of temperature. Ref.: invar and brass. [3.6×10^{-4} cm deg C^{-1}.]

7.8. A length measured with a travelling microscope is apparently 240.20 mm judged by the vernier scale at a temperature of 21°C. This scale is made of brass, and was calibrated accurately at 10°C. Calculate the true value of the measured length. Ref.: brass. [240.25 mm.]

7.9. A circular iron disc of area 5.0 cm² at a temperature of 0°C has exactly the same diameter as a hole in a brass plate at 0°C.

Calculate the width of the annular gap when the disc is set centrally in the hole and both metals are at a temperature of 85°C. Ref.: iron and brass. [7.5×10^{-4} cm.]

7.10. An aluminium ring is to be shrunk on to an iron rod of diameter 1.000 cm at 0°C. At this temperature the inner diameter of the ring is 20 μm less than the diameter of the rod. (a) To what temperature must the ring be heated to allow it to be slipped on to the rod with a radial clearance of 20 μm? (b) To what temperature must both the rod and the ring be heated after fitting, so that the ring is just loose? Ref.: aluminium and iron. [230°C; 140°C.]

7.11. A square sheet of steel has sides of length 20 cm at a temperature of 0°C. Calculate the increase of area at 40°C and the error in assuming that the coefficient of superficial expansion is twice that of linear expansion. Ref.: steel. [0.38 cm²; −0.024%.]

Volumetric Expansion

7.12. Worked Example

A mass of metal at a temperature of 200°C has a volume of 100 cm³. Calculate the increase of volume at 400°C assuming the coefficient of linear expansion to be 2.0×10^{-5} deg C^{-1} relative to (a) 0°C, and (b) 200°C.

Introduction

Consider a cube having sides of length l_1 at a temperature of θ_1 and l_2 at θ_2, where $\theta_2-\theta_1 = \Delta\theta$.

As in ex. 7.3,

$$l_2 = l_1[1+\alpha\Delta\theta]. \quad \therefore\ l_2^3 = l_1^3[1+\alpha\Delta\theta]^3.$$

Then, putting V for volume,

$$V_2 = V_1[1+\alpha\Delta\theta]^3 = V_1[1+3\alpha\Delta\theta+\ldots],$$

where the expansion includes terms involving $\alpha^2 \Delta\theta^2$ and $\alpha^3 \Delta\theta^3$, which, in most cases, are negligible in comparison with $\alpha\Delta\theta$.

Now if γ = coefficient of volumetric expansion,

$$V_2 = V_1(1+\gamma\Delta\theta). \tag{7.7}$$

Therefore $\gamma = 3\alpha$, to a fair approximation.

Alternatively, using ΔV for $V_2 - V_1$, (7.7) may be put into the same form as eqns. (7.2) and (7.6).

Thus
$$\Delta V = \gamma V \Delta\theta. \tag{7.8}$$

Solution

(a) When the coefficient is relative to 0°C, V_1 in eqn. (7.7), and V in eqn. (7.8) must be the volume when $\theta_1 = 0$°C.

Thus, using eqn. (7.7),

$$V_{200} = V_0[1+\gamma_0(200-0)],$$

where $\gamma_0 = 3\alpha_0 = 6.0+10^{-5}$ deg C^{-1}.

$$\therefore\ 100 = V_0[1+(6.0\times10^{-5}\times200)] = V_0\times1.012.$$
$$\therefore\ V_0 = 100/1.012 \text{ cm}^3.$$

Then
$$\begin{aligned} V_{400} &= V_0[1+\gamma_0(400-0)] \\ &= (100/1.012)[1+(6.0\times10^{-5}\times400)] \\ &= 102.4/1.012. \end{aligned}$$

$$\begin{aligned} \therefore \text{ expansion } V_{400}-V_{200} &= (102.4/1.012)-100 \\ &= 1.2/1.012 = 1.185 = 1.2 \text{ cm}^3 \quad \text{(2 sig. fig.)}. \end{aligned}$$

(b) Now assuming the coefficient to refer to 200°C,

$$\gamma_{200} = 3\times2.0\times10^{-5} \text{ deg C}^{-1},$$

and eqn. (7.8) gives

$$\Delta V = \gamma_{200}\times V_{200}\times(400-200) = 6.0\times10^{-5}\times100\times200 = 1.2 \text{ cm}^3.$$

Note that the closeness of the answers (a) and (b) shows that variation of γ with temperature may often be neglected, unless contrary information is given.

7.13. Worked Example

An aluminium flask has a mass of 20.00 g when empty, 52.11 g when completely filled with a liquid at 0°C, and 50.72 g when completely filled with the same liquid at 100°C. Calculate the real coefficient of volumetric expansion of the liquid. Ref.: aluminium.

Introduction

Let V_0 = volume of flask at 0°C = volume of liquid at 0°C.

Then expansion of flask $(f) = \Delta V_f = V_0 \gamma_f \Delta\theta$

and expansion of liquid $(l) = \Delta V_l = V_0 \gamma_l \Delta\theta$,

where, in each case, γ = real coefficient of volumetric expansion of the material concerned.

The apparent expansion of the liquid ΔV_a, as judged by the overflow, $= \Delta V_l - \Delta V_f = V_0(\gamma_l - \gamma_f)\Delta\theta$, where the part in brackets may be called the differential expansion coefficient or the apparent expansion coefficient γ_a. Then the apparent or differential expansion

$$\Delta V_a = V_0 \gamma_a \Delta\theta,$$

where

$$\gamma_a = \gamma_l - \gamma_f. \tag{7.9}$$

The coefficients of volumetric expansion may be used to calculate also the effect of temperature change on the density of a material. Thus when the volume of a constant mass m of a substance changes from V_1 to V_2, the density changes from $\varrho_1 = m/V_1$ to

$$\varrho_2 = m/V_2 = m/[V_1(1+\gamma\Delta\theta)].$$

$$\therefore \varrho_1 = \varrho_2(1+\gamma\Delta\theta). \tag{7.10}$$

Note that this equation has the suffixes on the opposite sides from that for V, eqn. (7.7).

Solution

The apparent expansion coefficient for the liquid in the flask is calculated from the change of the apparent density of the liquid. Thus, ignoring the expansion of the flask, it appears that the density of the liquid changes from

$$\varrho_0 = m_0/V \quad \text{to} \quad \varrho_{100} = m_{100}/V,$$

where V is the volume of the flask, and the mass of liquid in it changes from

$$m_0 = 52.11-20.00 \text{ g} \quad \text{to} \quad m_{100} = 50.72-20.00 \text{ g}.$$

Then substituting in eqn. (7.10) gives

$$32.11/V = (30.72/V)[1+\gamma_a(100-0)].$$

$$\therefore \ \gamma_a = (32.11-30.72)/(30.72\times 100)$$

$$= 1.39/3072 = 4.52+10^{-4} \text{ deg C}^{-1}.$$

Now the coefficient is corrected from the apparent to the real value, using the data of the reference section for aluminium: the coefficient of linear expansion

$$\alpha = 26\times 10^{-6} \text{ deg C}^{-1}.$$

$$\therefore \ \gamma_f = 3\alpha = 3\times 26\times 10^{-6} = 0.78\times 10^{-4} \text{ deg C}^{-1}.$$

Then, from eqn. (7.9),

$$\gamma_l = \gamma_a+\gamma_f = (4.52+0.78)\times 10^{-4} = 5.3\times 10^{-4} \text{ deg C}^{-1}.$$

7.14. Calculate the density of mercury at 150°C from the data of the reference section. [13.2×10^3 kg m^{-3}].

7.15. Calculate the increase of volume of a mass of 100 kg of steel when heated from 15°C to 70°C. Ref.: steel. [26 cm^3.]

7.16. A steel can is completely filled with 10 l. of paraffin oil at 0°C. Calculate the overflow when the temperature rises to 25°C. Ref.: steel and paraffin. [0.22 l.]

7.17. A cube of brass of 10 cm side is heated from 0°C to 50°C. Calculate the increase of volume and the error in assuming that the coefficient of volumetric expansion is three times that of linear expansion. Ref.: brass. [2.9 cm^3; -0.1%.]

7.18. A mercury-in-glass thermometer is to be constructed and filled with 1.0 cm^3 of mercury at 0°C. Calculate the diameter of the bore of the stem to give a sensitivity of 1.0 mm deg C^{-1}. Ref.: mercury and glass. [0.45 mm.]

7.19. Calculate the mass of a glass bottle filled with mercury at 0°C when the empty bottle has a mass of 20.0 g and the bottle filled at 100°C has a mass of 300.0 g. Ref.: glass and mercury. [304.4 g.]

7.20. A piece of glass weighs 50.00 g in air, and appears to weigh 31.50 g in water at 0°C, and 32.20 g in water at 100°C. Calculate the mean coefficient of volumetric expansion of the water. Ref.: glass. [4.2×10^{-4} deg C^{-1}.]

Thermal Stress

7.21. Worked Example

A steel rod of diameter 0.25 cm is clamped at its ends and has a tension of 750 N at a temperature of 20°C. Assuming that the length is maintained constant, calculate (a) the tension at 100°C, and (b) the temperature at which the tension would be zero. Ref.: steel.

Introduction

If the rod were free to expand or contract, an increase of temperature would cause an extension, given by eqn. (7.2): $\Delta l = \alpha l_0 \Delta\theta$, where l_0 = original length, α = temperature coefficient of linear expansion, and $\Delta\theta$ = rise of temperature. But the length is maintained constant: so a compressive strain is set up as if the rod were first allowed to expand and then compressed from the length $(l_0+\Delta l)$ back to the original length l_0. Then

$$\text{compressive strain} = (\text{decrease of length})/(\text{length before decrease})$$
$$= \Delta l/(l_0+\Delta l).$$

Now, since Δl is very small compared with l_0, it may be neglected in the denominator.

$$\therefore \text{ strain} = \Delta l/l_0 = \alpha l_0 \Delta\theta/l_0 = \alpha\Delta\theta.$$

But, as in ex. 6.5, stress $= E\times$strain, where $E =$ Young's modulus. Therefore the force due to thermal stress over the cross-section area A is given by

$$F = AE\alpha\Delta\theta. \tag{7.11}$$

Solution

For (a) the compressive force given by eqn. (7.11)

$$= (\pi 0.25^2\times 10^{-4}/4)(2.2\times 10^{11})(12\times 10^{-6})80 = 1040 \text{ N}.$$

But the rod had a tension of 750 N.

$$\therefore \text{ resultant compressive force} = 1040-750 = 290 \text{ N}.$$

This is a negative tension $= -290$ N.

In (b) the tension would be zero when the compressive force introduced just equalled the original tension of 750 N. But eqn (7.11) shows $F \propto \Delta\theta$ and from (a) $F = 1040$ N when $\Delta\theta =$ 80 deg C.

Therefore for (b)

$$\Delta\theta = (750/1040)80 = 58 \text{ deg C}.$$

Therefore the temperature for zero tension $= 20+58 = 78$°C

7.22. A steel bar of square cross-section 0.50×0.50 cm is heated from 10°C to 40°C. Calculate the force required to prevent change of length. Ref.: steel. [2.0×10^3 N.]

7.23. A rod of diameter 1.0 cm is fixed between rigid supports at its ends. Calculate the increase in tension when the temperature drops by 20 deg C. Ref.: steel. [4.1×10^3 N.]

7.24. A steel wire of diameter 0.50 mm is stretched and fixed at both ends under tension to a brass plate. The tension in the steel is 10 N at a temperature of 20°C. Calculate the tension at 5°C. Ref.: steel and brass. Assume negligible thermal stress in the brass. [5.8 N.]

CHAPTER 8

STATIC FLUIDS

Pressure in Fluids

8.1. Worked Example

A force of 5000 N is to be obtained from a hydraulic press having pistons of diameters 4.0 cm and 20 cm, operating at the same gravitational level. Calculate (a) the required pressure in the fluid, and (b) the force to be applied to the smaller piston.

Introduction

A substance is a perfect fluid if it will flow continually when a continual shearing force is applied. Thus air is a fluid since it has no rigidity or ability to hold itself in any particular shape against distorting shearing forces.

In a hydraulic press, shown in principle in Fig. 8.1, a distorting force can be applied to the fluid, which is a liquid, by depressing either piston. If a force F_1 is applied to the piston of area A_1,

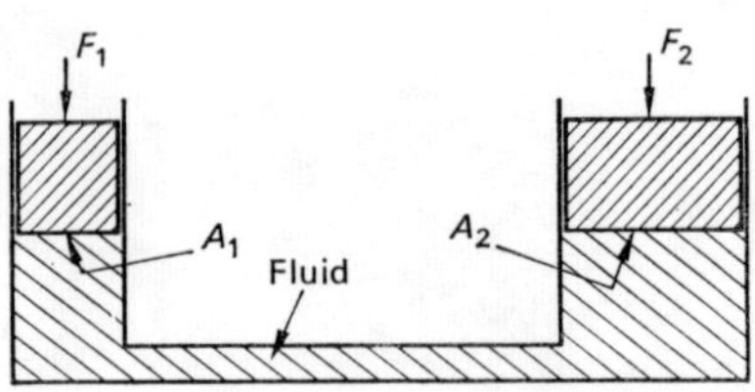

FIG. 8.1. Ex. 8.1.

all the fluid in contact with this area is affected equally and reacts with an opposing pressure $\delta F/\delta A$, where δF is the force of the fluid on any part δA of the area. When equilibrium is attained, the total force of the fluid on the piston balances F_1.

$$\therefore\ (\delta F/\delta A)A_1 = F_1, \quad \therefore\ \text{pressure } \delta F/\delta A = F_1/A_1. \tag{8.1}$$

δF must be perpendicular to δA, since a component of force along the boundary of the fluid cannot be produced without a shearing force in the fluid. Within the fluid, although not at the boundary, the pressure acts equally in all directions. This is shown by the equilibrium of a very small right-angled wedge of solid material, floating submerged at any inclination, as in Fig. 8.2.

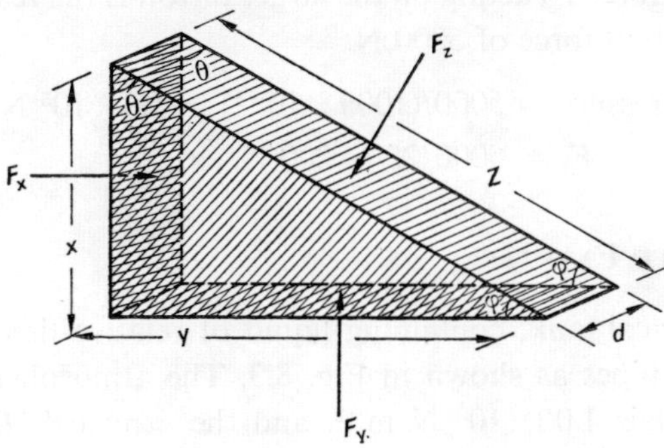

FIG. 8.2. Submerged wedge in equilibrium.

If all the angles are right angles except θ and φ in the triangular faces, and the forces shown are due to the pressures perpendicular to the rectangular faces, then, for equilibrium,

$$F_x = F_z \cos\theta.$$

Therefore pressure on face of area xd

$$F_x/xd = F_z \cos\theta/(zd\cos\theta) = F_z/zd.$$

Thus the pressures are equal in the directions of F_x and F_z whatever the value of θ. It follows that the pressure of the liquid perpendicular to any boundary is the same at all points at the same level. At different levels the pressures would be different due to the liquid having weight.

Solution

$$\text{Piston area } A_1 = \pi(4.0\times 10^{-2})^2/4 = 4\pi\times 10^{-4}\ \text{m}^2$$

and

$$A_2 = \pi(20\times 10^{-2})^2/4 = 100\pi\times 10^{-4}\ \text{m}^2.$$

Since the pressures on these areas are equal, and eqn. (8.1) applies to both,

$$\text{fluid pressure} = F_1/4\pi\times 10^{-4} = F_2/100\pi\times 10^{-4}.$$

But the force F_2 acting on the larger piston is the reaction equal to the required force of 5000 N.

$$\therefore\ \text{pressure} = 5000/(100\pi\times 10^{-4}) = 1.6\times 10^5\ \text{N m}^{-2} \quad \text{(a)}$$

and

$$F_1 = 5000/25 = 200\ \text{N}. \quad \text{(b)}$$

8.2. Worked Example

A spherical tank, containing liquid of relative density 0.90, is fitted with pipes as shown in Fig. 8.3. The atmospheric pressure acting at A is 1.00×10^5 N m^{-2}, and the stopcock D is closed. Calculate the pressures in the liquid at B and C, where $h_1 = 1.0$ m and $h_2 = 5.0$ m.

Introduction

In any body of liquid with a free surface, the pressure at any point depends upon the weight per unit volume, and upon the "head" of liquid, i.e. the vertical depth of the point below the free surface. If the surface has atmospheric or other pressure acting on it, this pressure is added to that due to the head of liquid. Thus, in the liquid at B, the pressure p_B is due to atmospheric pressure p_A acting on the free surface at A and pressure due to the liquid between A and B.

Now, if A = cross section area of tube, the volume of liquid between A and $B = h_1 A$. Also, if ϱ = density or mass per unit volume of liquid, and g = acceleration of gravity, then ϱg = weight per unit volume of liquid. Therefore weight of liquid in tube

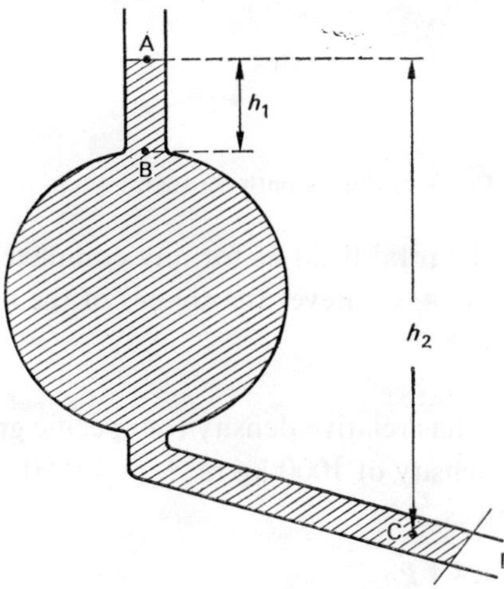

FIG. 8.3. Ex. 8.2.

$AB = h_1 A \varrho g$, in absolute units, and the total downward force acting on the liquid at $B = h_1 A \varrho g + p_A A$. This is supported entirely by the pressure p_B acting upward on the area A at B.

$$\therefore \; p_B A = h_1 A \varrho g + p_A A$$

$$\therefore \; p_B = h_1 \varrho g + p_A, \tag{8.2}$$

where $h_1 \varrho g$ is the pressure component due to the liquid.

The pressure at C is given similarly by substituting the head h_2, shown in Fig. 8.3, in place of h_1. The fact that there is no need for a straight vertical run between C and the surface is seen by considering a pipe fitted vertically upward from C and filled to the same level as A. This would not alter the pressure anywhere in the liquid.

Alternatively, a continuous zigzag path may be considered between C and A, shown partially in Fig. 8.4. The horizontal portions of such a path do not affect the pressure difference since pressures are equal at the same level. Thus the vertical portions

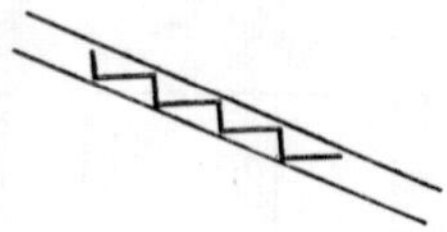

FIG. 8.4. Zigzag path through liquid.

would add up to the total head h_2 for any complete path through the liquid from C to A, whatever the shapes of the tank and pipes.

Solution

Since the liquid has relative density (or specific gravity) of 0.90, and water has a density of 1000 kg m^{-3}, $\varrho = 0.90\times 1000$ kg m^{-3}. Then eqn. (8.2) gives,

$$\begin{aligned} p_B &= h_1\varrho g + p_A \\ &= (1.0\times 0.90\times 1000\times 9.81)+(1.00\times 10^5) \\ &= (8.8\times 10^3)+(1.00\times 10^5)\ \text{N m}^{-2}, \end{aligned}$$

i.e. p_B exceeds the atmospheric pressure by 8.8×10^3 N m^{-2} and has a total value of 1.09×10^5 N m^{-2} (3 sig. fig.).

At C, $h_2 = 5.0$ m. Therefore p_C exceeds atmospheric pressure by $5.0\times 8.8\times 10^3 = 44\times 10^3$ N m^{-2}, and has a total value of 1.44×10^5 N m^{-2}.

8.3. A cylindrical tank of internal diameter 20 cm and height 15 cm has a pipe of internal diameter 3.0 cm set axially in one end and running 15 cm vertically upward from the tank. Water is poured in to fill (a) the tank only, and (b) the tank and the pipe. In each case, calculate the total weight of water and the pressure on the base of the tank. Ref.: water. [(a) 46 N, 1.5×10^3 N m^{-2}; (b) 47 N, 2.9×10^3 N m^{-2}.]

8.4. A cylindrical jar of diameter 5.0 cm contains mercury to a depth of 2.0 cm. Calculate the depth of water required in another jar of internal diameter 15 cm to make (a) equal weights of the two liquids, and (b) equal pressures on the bases of the respective jars. Ref.: water and mercury. [(a) 3.0 cm; (b) 27 cm.]

8.5. A tank has vertical sides and a horizontal base of internal area 25 m^2. A gauge, operated by the pressure at the base, is used to indicate the level of oil in the tank. Full-scale deflection of the gauge is given by an oil of relative density 0.90 for a depth of 5.0 m. Calculate (a) the depth of another oil of relative density 0.80, which would give full-scale deflection of the gauge, and (b) the weight of this quantity of oil. Ref.: water. [(a) 5.6 m; (b) 1.1×10^5 kg.]

8.6. A hydraulic jack, containing oil of relative density 0.90, is used to raise a load of 1.0 tonne. The smaller piston is operated by a lever to which the piston and handle are connected so that the respective forces are parallel to each other, perpendicular to the lever, and at distances of 5.0 cm and 50 cm respectively from the fulcrum. The pistons have diameters of 2.0 cm and 10 cm respectively. Assuming 100% efficiency, calculate (a) the force required on the handle, (b) the velocity ratio, and (c) the head of oil which could give the required pressure. Ref.: water. [(a) 39 N; (b) 250; (c) 140 m.]

Archimedes' Principle

8.7. Worked Example

A piece of brass weighs 25.0 gf in air and 22.1 gf when fully immersed in water at 0°C. When a piece of cork of density 250 kg m^{-3} is tied to the brass, they float together, fully submerged in the water. Calculate (a) the density of the brass, and (b) the weight of the cork alone in air.

Introduction

When the brass is suspended in the water, the apparent weight is less than in air because of the upthrust exerted by the water. This upthrust is due to the pressure of the surrounding water, being greater at the lower level below the brass than at the higher level above it.

If the brass were removed and its place filled by additional water, this water would be supported in equilibrium by the same upthrust since the surrounding water would be unchanged (Fig. 8.5). Therefore the upthrust in both cases would equal the weight of water in the same volume as the brass.

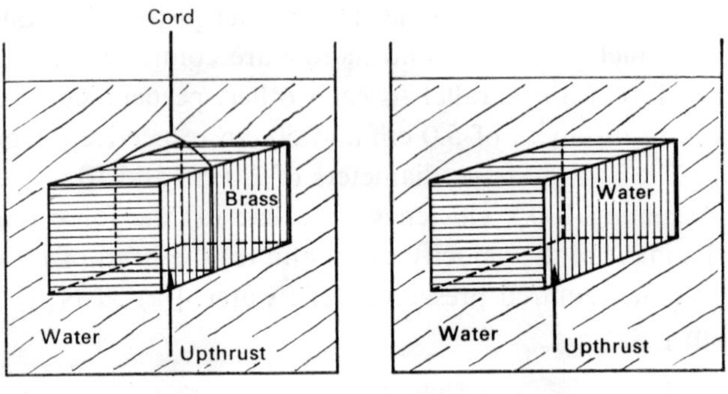

FIG. 8.5. Archimedes' principle (ex. 8.7).

Similarly, if the cork is submerged, the upthrust of the water on the cork equals the weight of water displaced by the cork. This upthrust is greater than the weight of the cork, since water is more dense than cork. Thus the excess upthrust tends to cause the cork to rise to the surface, but in the question this excess is used to support the remaining weight of the brass.

Not only water, but all fluids including air, exert upthrusts of this kind. Thus objects appear to weigh less in air than in vacuum.

In general, Archimedes' principle may be stated:

When a body is wholly or partially immersed in a fluid, it experiences an upthrust equal to the weight of displaced fluid.

Solution

(a) Apparent loss of weight of brass in water $= 25.0 - 22.1 = 2.9$ gf,

$$\therefore \text{ weight of water displaced by brass} = 2.9 \text{ gf}.$$

But weight per unit volume of water $= 1000$ kgf m^{-3},

$\therefore$ volume of water equal to brass $= 2.9 \times 10^{-3}/1000 = 2.9 \times 10^{-6}$ m^3.

Then density of brass = mass/volume

$$= 25.0 \times 10^{-3}/(2.9 \times 10^{-6}) = 8.6 \times 10^3 \text{ kg m}^{-3}. \qquad \text{(a)}$$

(b) If V = volume of cork in m^3, weight of cork $= V \times 250$ kgf and weight of water displaced by cork $= V \times 1000$ kgf.

$$\therefore \text{ excess upthrust} = V \times 750 \text{ kgf}.$$

This must support the remaining weight of the brass, i.e. 22.1×10^{-3} kgf.

$$\therefore V = 22.1 \times 10^{-3}/750 = 29.5 \times 10^{-6} \text{ m}^3.$$

$\therefore$ weight of cork alone $= 29.5 \times 10^{-6} \times 250 = 7.4 \times 10^{-3}$ kgf. (b)

8.8. Worked Example

A common hydrometer of constant weight floats in water with 2.0 cm of its stem above the water. When it is placed in a solution of relative density 1.20 the exposed length of stem is 10.0 cm. Assuming the stem to be of constant cross-section, calculate the relative density of the liquid when the exposed length is 6.0 cm.

Introduction

The common hydrometer, shown in Fig. 8.6, has a graduated glass stem rising from a hollow glass float weighted at the bottom. It sinks in a liquid, e.g. acid or spirit, to a depth which depends upon the density of the liquid. Following Archimedes' principle,

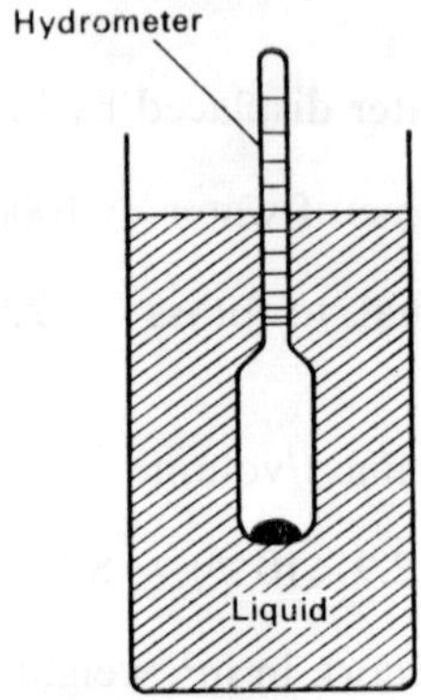

FIG. 8.6. The hydrometer (ex. 8.8).

any object, such as the hydrometer, which floats freely in equilibrium is supported by the upthrust which would support the displaced liquid.

If the hydrometer is pressed down and released, it is forced up again by the increased upthrust equal to the weight of the increased quantity of displaced fluid. Similarly, it returns to the same equilibrium level if raised and released. However, if the density of the liquid increases, the hydrometer rises to a new equilibrium position since it displaces a smaller volume for the same weight. Hence the stem may be calibrated to indicate the density of the liquid, although the scale is not linear.

If V is the total volume of the hydrometer, and A and l are the cross-section area and exposed length of the stem, the submerged volume $= V - lA$.

$$\therefore \text{ weight of liquid displaced} = \varrho(V - lA),$$

where ϱ is the density of liquid. Then, for the equilibrium position,

$$\varrho(V-lA) = \text{weight of hydrometer.} \tag{8.3}$$

Note that l depends upon ϱ but is not proportional to ϱ in this equation.

Solution

When $\varrho = 1000$ kg m^{-3} for water, and $l = 0.020$ m, eqn. (8.3) gives

$$1000(V-0.020A) = \text{weight } w. \tag{8.4}$$

Similarly, when $\varrho = 1.2\times 1000$ kg m^{-3} and $l = 0.100$ m, eqn. (8.3) gives

$$1200(V-0.100A) = w. \tag{8.5}$$

Finally, in the third instance,

$$\varrho_3(V-0.060A) = w, \tag{8.6}$$

where ϱ_3 is the unknown density.

Now to solve for ϱ, V and A are first found in terms of w.

Thus, eqn. (8.4)×6: $\quad 6000V-120A = 6w;$ (8.7)

eqn. (8.5): $\quad 1200V-120A = w.$ (8.8)

Subtraction gives: $\quad 4800V = 5w.$

$$\therefore\ V = 5w/4800 = 1.042\times 10^{-3}w.$$

$\therefore$ eqn. (8.4) becomes $1.042w-20A = w. \quad \therefore\ A = 2.1\times 10^{-3}w.$

Then eqn. (8.6) becomes

$$\varrho_3[(1.042\times 10^{-3}w)-(0.060\times 2.1\times 10^{-3}w)] = w.$$

$$\therefore\ \varrho_3 = 10^3/(1.042-0.126) = 1090 \text{ kg m}^{-3}.$$

Therefore the density relative to water is 1.09.

8.9. A piece of nickel weighs 6.60 gf in air, 5.85 gf when submerged in water, and 6.00 gf when submerged in paraffin oil.

Calculate the densities of nickel and paraffin oil. Ref.: water. [8800 kg m^{-3}; 800 kg m^{-3}.]

8.10. To determine the density of cork, a piece which weighs 10 gf in air is tied to a piece of metal. When suspended in water, the cork and metal together have an apparent weight of 15 gf and the metal alone 45 gf. Calculate the density of the cork. [250 kg m^{-3}.]

8.11. A sealed bulb, of weight 22.0 gf in air, is made of glass of density 2500 kg m^{-3}. When totally immersed in water the apparent weight is 10.0 gf. Calculate the internal and external volumes of the bulb, neglecting the density of air. [3.2 cm^3; 12 cm^3.]

8.12. A sample of solder is known to be an alloy of tin of density 7300 kg m^{-3} and lead of density 11300 kg m^{-3}. A piece of this solder weighs 300 gf in air and 265 gf in water. Calculate the proportions by weight of tin and lead, assuming that the volume of the solder equals the sum of the volumes of the tin and lead separately. Ref.: water. [58/42.]

8.13. The fabric of a balloon has a mass of 10 kg and mean density of 1000 kg m^{-3}. The balloon may be inflated to enclose a volume of 20 m^3. Calculate the maximum load that may be carried when the balloon is inflated with hydrogen of density 0.090 kg m^{-3} in a surrounding atmosphere of density 1.2 kg m^{-3}. [12 kg.]

8.14. The density of a liquid is determined by weighing a known volume of the liquid against brass balance weights of density 8500 kg m^{-3} in air of density 1.29 kg m^{-3}. Calculate the percentage error due to ignoring the density of air, when the density of the liquid is 790 kg m^{-3}. [−0.15%.]

Fortin Barometer

8.15. Worked Example

A Fortin barometer indicates an atmospheric pressure of 770 mmHg. Calculate (a) the pressure in absolute units, and (b) the head of water which would give the same pressure.

Introduction

Atmospheric pressure at any altitude is due primarily to the weight of air above that level, although it is complicated by movement of air, and by a variation of density with altitude.

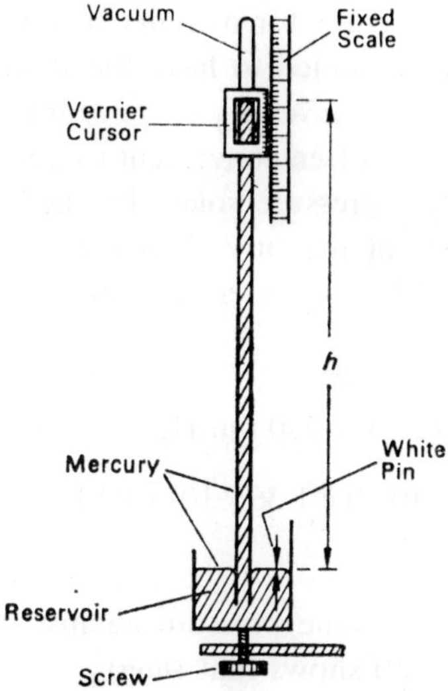

FIG. 8.7. The Fortin barometer (ex. 8.15).

In a Fortin barometer, shown in essence in Fig. 8.7, atmospheric pressure acts freely on the surface of the mercury in the reservoir, whereas only the negligible pressure of mercury vapour and unavoidable traces of air act on the mercury surface under the vacuum in the tube. Then the atmospheric pressure is balanced by the pressure due to the mercury column. The height h of this column is measured by first setting the reservoir level so that the image of the white pin in the mercury surface just touches the tip of the pin itself, fixed at the zero level of the scale, and then reading the level of the vernier cursor when it is set at the level of mercury in the tube.

The mercury pressure at zero level in the tube is given by $p = h\varrho g$, as included in eqn. (8.2). This equals the atmospheric pressure p_A at the same level on the open surface in the reservoir.

$$\therefore\ p_A = h\varrho g, \tag{8.9}$$

where $\varrho = 1.36\times10^4\,\text{kg}\,\text{m}^{-3}$ for mercury at 0°C and, unless stated otherwise, g may be assumed to have the standard value of 9.81 m s^{-2}. Then eqn. (8.9) gives p_A in newtons per square metre (N m^{-2}). However, it is often convenient to quote the atmospheric pressure or other gas pressure solely by the value of h, usually stated in millimetres of mercury (mmHg). The standard atmospheric pressure is 760 mmHg, equivalent to 1.013×10^5 N m^{-2}.

Solution

(a) From eqn. (8.9) for 770 mmHg,

$$p_A = h\varrho g = (770\times10^{-3})\times1.36\times10^4\times9.81 = 1.03\times10^5\ \text{N m}^{-2}. \tag{a}$$

(b) To balance the same pressure against mercury (Hg) and water (H_2O), eqn. (8.9) shows that $(h\varrho g)_{Hg} = (h\varrho g)_{H_2O}$.

$$\therefore\ (h)_{H_2O} = 770\times10^{-3}\times1.36\times10^4/10^3 = 10.5\ \text{m}. \tag{b}$$

8.16. A barometer indicates a pressure of 660 mmHg at the top of a mountain when the pressure at sea level is 760 mmHg. Assuming that the mean density of air between sea level and mountain top is 1.2 kg m^{-3}, calculate (a) the height of the mountain, and (b) the change of barometer indication due to a change of altitude of 10 m. Ref.: mercury. [(a) 1.1 km; (b) 0.9 mmHg.]

8.17. Calculate the correction factor required to convert the readings of a Fortin barometer at a temperature of 20°C to the equivalent values at 0°C. Assume that the scale is engraved on brass which runs continuously down to zero level. Ref.: brass and mercury. [0.9968.]

U-tube Manometers

8.18. Worked Example

An open U-tube manometer indicates a gauge pressure of 150 mmHg when the absolute gas pressure is 1.24×10^5 N m^{-2}. Calculate (a) the atmospheric pressure, and (b) the absolute gas pressure when the indication is 200 mmHg. Ref.: density of mercury.

Introduction

A manometer is a device used in measuring the pressure of gas in a container which is connected to it. One side of a U-tube, containing a liquid, is connected to the container, while the other side may be closed or open to the atmosphere as in Fig. 8.8.

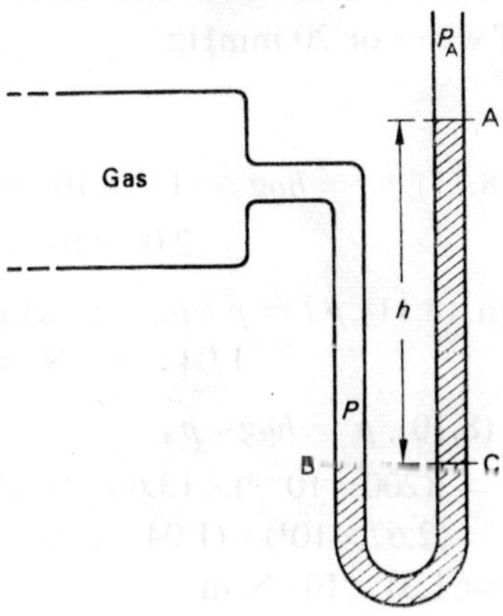

FIG. 8.8. The manometer (ex. 8.18).

If the gas pressure on the liquid surface at B is greater than the atmospheric pressure at A, then in equilibrium the level A is higher than B. Assuming negligible curvature of the liquid surfaces, the pressure in the liquid at the surface B equals the gas pressure p

and equals the pressure at the same level C in the other branch. The pressure at C may be attributed to the liquid and atmosphere directly above it. Therefore, as in eqn. (8.2),

$$p = h\varrho g + p_A, \tag{8.10}$$

where h is the difference of levels A and B, ϱ is the density of the liquid in the tube, g is the acceleration of gravity, p_A is the atmospheric pressure, both p and p_A being in units of force per unit area.

In many cases the important quantity is not the absolute pressure p but the amount by which p exceeds p_A. This is called the "gauge pressure" p_G because the gauge is operated by this difference and indicates it by the height h. Thus

$$p_G = p - p_A = h\varrho g. \tag{8.11}$$

Frequently, p, p_A, or p_G may be stated in terms of height, as for the barometer of ex. 8.15, although it is necessary to state the liquid also, e.g. 27 cm of water or 20 mmHg.

Solution

(a) From eqn. (8.11), $p_G = h\varrho g = (150\times10^{-3})\times13.6\times10^3\times9.81$
$= 2.00\times10^4 \text{ N m}^{-2}$.

Then, also from eqn. (8.11), $p_A = p - p_G = (1.24\times10^5) - (2.00\times10^4)$
$= 1.04\times10^5 \text{ N m}^{-2}$. (a)

(b) From eqn. (8.10), $p = h\varrho g + p_A$
$= [(200\times10^{-3})\times13.6\times10^3\times9.81] + (1.04\times10^5)$
$= (2.67\times10^4) + (1.04\times10^5)$
$= 1.31\times10^5 \text{ N m}^{-2}$. (b)

8.19. Worked Example

An open tank containing oil of relative density 0.90 has a mercury manometer used as a level indicator. The zero of the scale is at the same level as the base of the tank. Calculate (a) the depth of oil in the tank when the upper mercury level is 12 cm above zero, and (b) the corresponding pressure of oil on the base of the tank.

Introduction

Since the pressure in the oil near the base of the tank depends upon the depth of the oil, the pressure may give an indication of the depth. For this purpose, the pressure need not be determined in absolute units and may be indicated conveniently by a U-tube manometer or other pressure gauge. If the scale of the manometer is calibrated directly in terms of depth of oil, the indication is correct only for oil of one density.

Figure 8.9 shows a U-tube manometer of constant bore diameter such that *C* falls and *A* rises by equal amounts when oil

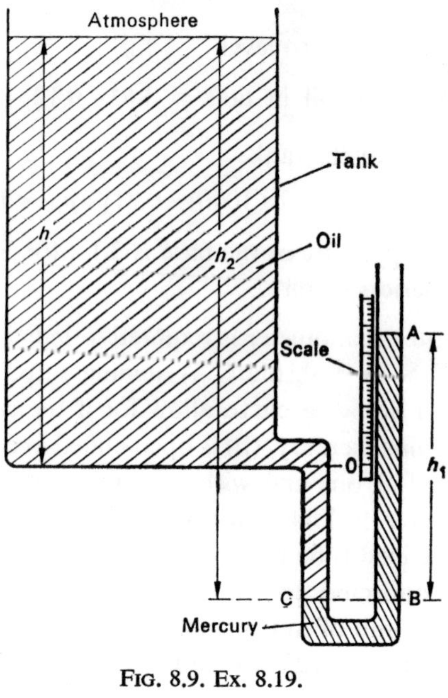

FIG. 8.9. Ex. 8.19.

enters the tank. Then the mercury column AB has height h_1, equal to twice the height on the scale, which has its zero at the base level of the tank.

Now the pressure at level B in the mercury $= p_B = h_1\varrho_1 g + p_A$, and the equal pressure at $C = h_2\varrho_2 g + p_A$, where suffix 1 is for mercury and 2 for oil.

$$\therefore\ h_1\varrho_1 g + p_A = h_2\varrho_2 g + p_A.$$

$$\therefore\ h_1\varrho_1 = h_2\varrho_2. \tag{8.12}$$

Solution

(a) Since A is 12 cm above zero level, $h_1 = 0.24$ m. Also, since the oil has relative density (or specific gravity) of 0.90, and water has density of 1000 kg m^{-3}, ϱ_2 for oil $= 0.90 \times 1000$ kg m^{-3}.

Then, from eqn. (8.12),

$$h_2 = h_1\varrho_1/\varrho_2 = 0.24 \times 1.36 \times 10^4/(0.90 \times 1000) = 3.63 \text{ m.}$$

But C is 0.12 m below zero level.

$$\therefore\ \text{depth } h \text{ of oil in tank} = 3.63 - 0.12 = 3.5 \text{ m.} \qquad \text{(a)}$$

(b) At the base of the tank,

$$p = h\varrho g + p_A = [3.5 \times (0.90 \times 1000) \times 9.81] + p_A = (3.1 \times 10^4) + p_A.$$

Therefore the pressure at the base is 3.1×10^4 N m^{-2} in excess of the atmospheric pressure. (b)

8.20. A U-tube manometer is required to indicate gas pressures up to 1.6×10^5 N m^{-2} absolute. Assuming s.t.p. (ref.), calculate the difference of liquid levels to allow for if the tube (a) is open-ended and contains mercury, (b) is open-ended and contains water, and (c) is closed at one end with a vacuum over mercury. Ref.: mercury and water. [(a) 440 mm; (b) 6.0 m; (c) 1.2 m.]

8.21. An inclined-tube manometer has a vertical limb of large cross-section connected to the gas under measurement and a straight inclined limb of much smaller section open to the atmosphere. The inclination is 25° to the horizontal, and the liquid used is an oil of density 0.90 g cm^{-3}. Neglecting the change of level in the wide limb, calculate the change of gas pressure when the level moves 10 cm along the inclined limb. Give the answer in mmHg as if measured by a vertical U-tube. Ref.: mercury. [2.8 mmHg.]

8.22. An open-ended U-tube manometer for measuring steam pressure contains an oil of density 900 kg m^{-3} and has a column of water 5.0 cm high condensed on the oil in the tube containing steam. Calculate the steam pressure when the oil level in the open tube is 15.0 cm higher than that below the water. [8.3×10^2 N m^{-2}, gauge pressure.]

8.23. A cistern manometer has a single vertical tube of bore diameter 5.0 mm standing in a cylindrical cistern or tank in which the liquid surface has a diameter of 5.0 cm. If the pressure on the liquid surface in the cistern is to be measured by the rise of liquid in the tube, calculate the error in neglecting the fall of level in the cistern. [-1%.]

8.24. The temperature of a mercury manometer rises by 20°C while the pressures of gas and atmosphere remain constant. Calculate the change of reading relative to the original. Ref.: mercury. [$+0.36$%.]

Surface Tension

8.25. Worked Example

A soap bubble in air has a diameter of 5.0 cm and the solution has a surface tension of 0.030 N m^{-1}. Calculate (a) the excess pressure inside the bubble, and (b) the work done against surface tension in blowing the bubble at constant temperature.

Introduction

All molecules in a liquid are subject to forces of attraction and repulsion relative to their neighbours. Both types of force decrease with increase of separation of the molecules, but the repulsive forces decrease more rapidly than the attractive. Thus if the separation of two molecules increases beyond the equilibrium value, the resultant force is attractive, and work must be done to make such an increase. There are several important consequences:

(1) Molecules which move slightly out of line from the surface of a liquid, due to thermal motion, receive a resultant force which tends to draw them back. The net result of such forces at the surface is described as surface tension, as if the surface molecules formed an elastic skin. A measure of this is derived from Fig. 8.10

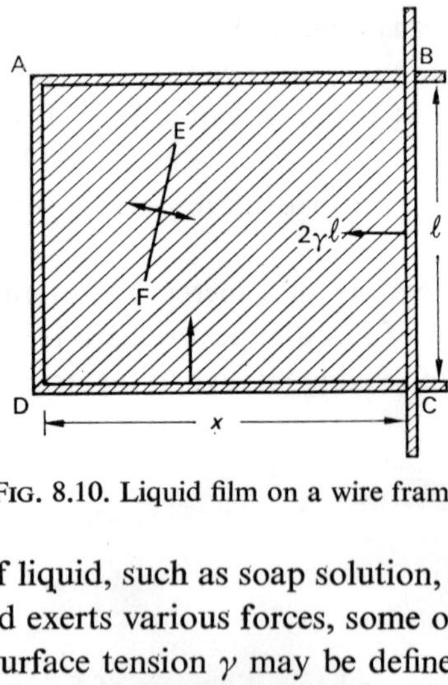

FIG. 8.10. Liquid film on a wire frame.

in which a film of liquid, such as soap solution, is spread in a wire frame *ABCD*, and exerts various forces, some of which are shown by arrows. The surface tension γ may be defined as the force per unit length exerted by a liquid surface as on the wire *BC*. Since the film has two surfaces, front and back,

$$\text{the total force on } BC \text{ is } 2\gamma l. \tag{8.13}$$

Similarly, there is a force on each side of a line such as *EF* drawn anywhere on the surface. If a cut could be made along *EF*, such forces would cause the liquid to fly back to the wire.

(2) When the surface area of a liquid is increased, as when a soap bubble is blown, work must be done to increase the separa-

tion of some surface molecules so that others may become exposed or may reach the surface from the interior of the liquid. The energy supplied for this purpose is described as free surface energy, and may be recovered when the area decreases. Thus if the wire *BC* of Fig. 8.10 is drawn a distance x from *AD* so as to spread the film, the work done against the force of (8.13) is $2\gamma lx$. Then the free surface energy per unit area of each surface

$$= (2\gamma lx)/(2lx) = \gamma. \tag{8.14}$$

Thus the energy per unit area equals the force per unit length: in M.K.S. units, J m^{-2} is equivalent to N m^{-1}.

(3) Molecules, which may move in any direction within the liquid, are more restricted and retarded on arriving at the surface, and so lose thermal kinetic energy. Therefore, when the surface area increases and more molecules lose such energy, the temperature falls unless heat is supplied.

(4) The surface tension around a spherical drop of liquid, or around a gas bubble in a liquid, causes the internal pressure to exceed the external pressure by an amount Δp, found as follows. If the radius of the drop or bubble were to increase slightly by δr, the work done in this movement by the total outward force $\Delta p 4\pi r^2$ would provide the free surface energy $\gamma \delta A$ for the extra area δA.

$$\therefore \quad \Delta p 4\pi r^2 \, \delta r = \gamma \delta A.$$

But $A = 4\pi r^2$, and differentiation gives $\delta A/\delta r = 8\pi r$ for very small increments.

$$\therefore \quad \Delta p 4\pi r^2/\gamma = 8\pi r.$$

$$\therefore \quad \Delta p = 2\gamma/r \tag{8.15}$$

for a single spherical surface. Whatever the curvature of the surface, the pressure is greater on the concave side.

In the case of a spherical soap bubble with air on both sides, eqn. (8.15) applies to both inner and outer surfaces. Thus the air pressure inside the bubble exceeds the pressure in the soap film by Δp, and the pressure in the soap film exceeds the external air pressure also by Δp, making a difference of air pressures of $2\Delta p$.

Solution

(a) For the double surface of the bubble, $2\Delta p = 4\gamma/r$

$$= 4\times 0.030/(2.5\times 10^{-2}) = 4.8 \text{ N m}^{-2}. \qquad \text{(a)}$$

(b) Work done against surface tension

$$= \gamma\times(\text{total area of inside and outside}) = \gamma\times 8\pi r^2$$

$$= 0.030\times 8\pi\times 6.25\times 10^{-4} = 4.7\times 10^{-4} \text{ J}. \qquad \text{(b)}$$

Note that additional work must be done to build up the air pressure inside the bubble and to force back the external atmosphere as the bubble grows; but the question does not ask for this to be calculated.

8.26. Worked Example

The tube of a Fortin barometer has a bore of 5.0 mm diameter. Calculate the error of indication if no allowance is made for surface tension. The surface tension of mercury is 0.45 N m^{-1} with angle of contact of 140°.

Introduction

At the boundary of a liquid and a solid, the resultant surface tension depends not only upon the forces between molecules of the liquid but also upon forces between molecules of liquid and solid. Thus, due to differences of these forces, water completely wets perfectly clean glass, whereas mercury tends to gather itself away from glass. Thus water runs up parallel to the surface of the bore of the tube of Fig. 8.11(a), whereas mercury retreats down the tube of Fig. 8.11(b).

In both cases (a) and (b) the difference of levels inside and outside the tube is related to the pressure difference Δp across the

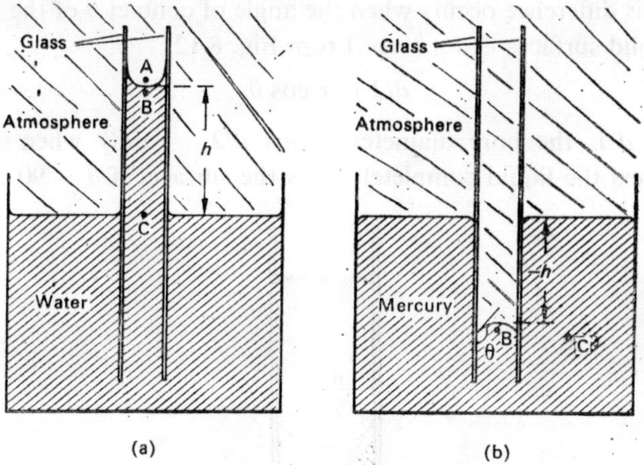

FIG. 8.11. (a) Capillary rise of water in a tube. (b) Capillary fall of mercury in a tube.

curved surface, or meniscus, of the liquid in the tube. Thus in (a) h is the "capillary rise" such that the pressure changes down the tube from atmospheric p_A at A to $p_A - \Delta p$ at B just below the surface, and to

$$p_A - \Delta p + h\varrho g = p_A \tag{8.16}$$

at the level C equal to that of the surrounding flat surface.

In case (b) the capillary "rise" is negative, and Δp is negative in comparison with (a) because the meniscus is convex. Then, for equal pressures at B and C in case (b),

$$p_A - \Delta p = p_A - h\varrho g, \tag{8.17}$$

where both Δp and h are numerically negative.

Both eqns. (8.16) and (8.17) give $h\varrho g = \Delta p$, where Δp is given by eqn. (8.15).

$$\therefore \; h\varrho g = \Delta p = 2\gamma/r. \tag{8.18}$$

Then h is considerable when r is very small. Here the radius of curvature r of the meniscus is assumed to be constant, but may be different from the radius of the bore of the tube.

This difference occurs when the angle of contact θ of the liquid and solid surfaces is not zero. From Fig. 8.12,

$$d/2 = r\cos\theta, \tag{8.19}$$

where d is the bore diameter. Thus $d/2 = r$ only when $\theta = 0$, i.e. when the liquid completely wets the surface. If $\theta > 90°$, as in

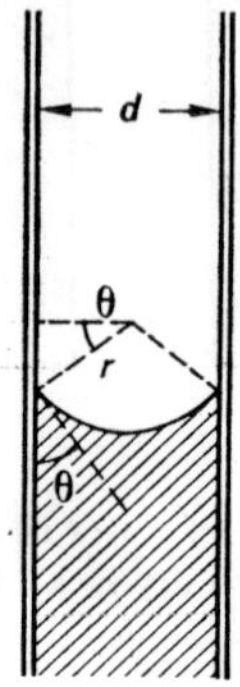

FIG. 8.12. Angle of liquid contact.

Fig. 8.11(b) for mercury in glass, $\cos\theta$ is negative, and so also are r and h.

In a barometer tube the mercury stands higher than the surrounding mercury level, but the height is less if the tube is narrow than if it is wide, due to $\Delta p \propto 1/r$.

Solution

In the particular tube of mercury, $d/2 = 2.5\times10^{-3}$ m and

$$\cos\theta = \cos 140° = -\cos 40° = -0.766.$$

Then, from eqn. (8.19),

$$r = -2.5\times10^{-3}/0.766 = -3.27\times10^{-3} \text{ m}.$$

So eqn. (8.18) gives

$$h = 2\gamma/r\varrho g = -2\times0.45/(3.27\times10^{-3}\times1.36\times10^{4}\times9.81)$$
$$= -2.1\times10^{-3} \text{ m}.$$

This is the depression of the mercury level due to Δp, and is therefore the amount by which the barometer height falls short of the full value which would occur in a very wide tube.

8.27. When a capillary tube of bore diameter 0.50 mm is dipped vertically into a vessel of liquid of density 0.70 g cm^{-3}, the liquid rises in the tube to a height 3.5 cm above that in the vessel. Calculate (a) the difference of pressures at points just inside and just outside the surface of the liquid in the tube, and (b) the surface tension of the liquid if it has an angle of contact of 40°. [(a) 240 N m^{-2}; (b) 0.039 N m^{-1}.]

8.28. A glass tube of bore diameter 0.20 mm stands vertically in a beaker of water of surface tension 0.073 N m^{-1}. Calculate (a) the capillary rise, assuming that the tube is of adequate length, and (b) the radius of curvature of the meniscus at the top of the tube if this is cut square at a height of 10 cm above the level of water in the beaker. [(a) 15 cm; (b) 0.15 mm.]

8.29. A common hydrometer with cylindrical stem of 2.0 mm diameter floats in liquid which wets the stem completely and has a relative density of 1.2 and surface tension of 0.030 N m^{-1}. Calculate the effect of the surface tension on the depth to which the hydrometer sinks. [Depth increased by 5.1 mm.]

8.30. A rectangular glass slide is 8.0 cm long and 2.0 mm thick. It is suspended, dipping vertically into water, with its long edge horizontal and 0.50 cm below the surface. If the apparent weight is then 0.40 gf greater than in air, calculate the surface tension of water. [0.072 N m^{-1}.]

8.31. Liquid is at rest in a vertical U-tube with open ends. The difference of liquid levels in the two limbs, which have bore diameters of 0.50 mm and 0.80 mm respectively, is 6.5 mm. Calculate the surface tension of the liquid assuming that it completely wets the surface of the tube and has a relative density of 0.85. [18×10^{-3} N m^{-1}.]

CHAPTER 9

IDEAL GASES

Constant Temperature

9.1. Worked Example

A quantity of coal gas occupies a volume of 1.6 l. at a gauge pressure of 1000 mmHg. Calculate the gauge pressure when the volume is expanded to 6.4 l. Assume constant temperature and an atmospheric pressure of 760 mmHg.

Introduction

Boyle's law states that the volume of a fixed mass of gas at constant temperature is inversely proportional to the pressure. Thus for constant mass and temperature, if V = volume and p = pressure,

$$V \propto 1/p, \quad \text{or} \quad (pV) \text{ is constant.} \tag{9.1}$$

Figure 9.1 shows several rectangles of equal areas, representing the constant product $(pV) = 8$ in unspecified units for a particularly simple case. This shows that if p is doubled, V is halved, and vice versa, under constant temperature (isothermal) conditions.

Real gases do not follow Boyle's law exactly, particularly at temperatures and pressures which bring them near to changing into liquids. However, real gases follow the ideal of Boyle's law to sufficient accuracy for many purposes, and this may be assumed unless there is clear evidence to the contrary, as with saturated vapour (see ex. 15.8).

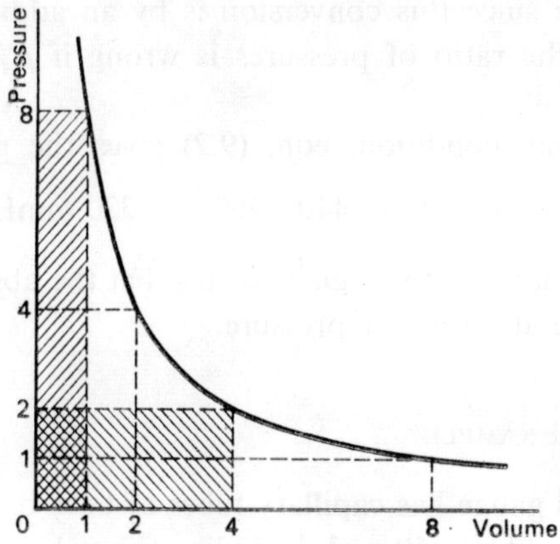

FIG. 9.1. Constant product curve for pressure and volume.

The above pressure p is the absolute pressure as distinct from the differential pressure or gauge pressure p_G, which is the amount by which the absolute pressure p exceeds the atmospheric pressure p_A. Then, as in ex. 8.19,

$$p = p_G + p_A. \tag{9.2}$$

Solution

Using eqn. (9.2), the initial pressure $p_1 = 1000+760 = 1760$ mmHg. Then, from (9.1), $p_1V_1 = p_2V_2$, where $V_1 = 1.6$ l. and $V_2 = 6.4$ l.

$$\therefore \text{ the final pressure } p_2 = p_1V_1/V_2$$
$$= 1760 \times 1.6/6.4 = 440 \text{ mmHg.}$$

Note that the mixture of units involving both millimetres and litres is allowable, since the calculation is one of proportion. Thus if the volumes were converted to cubic metres and the pressures to newtons per square metre, V_2/V_1 and p_1/p_2 would still equal 4. However, the conversion from gauge pressure to absolute pressure must be made since this conversion is by an addition, not by a factor. Thus the ratio of pressures is wrong if p_A is not added to p_G.

For the final condition, eqn. (9.2) gives the gauge pressure

$$p_G = p_2 - p_A = 440 - 760 = -320 \text{ mmHg}.$$

Here the final negative sign indicates that the absolute pressure is less than the atmospheric pressure.

9.2. Worked Example

A McLeod gauge has capillary tubes of bore diameter 1.0 mm and a bulb of volume 50 cm³, including the volume of the tubing in which gas is trapped. When the mercury in the bypass capillary of the main tube is level with the top of the closed capillary, the length of the trapped gas column is 20 mm. Calculate the original gas pressure under measurement.

Introduction

A McLeod gauge is essentially a manometer for the measurement of absolute pressures of gas from a few millimetres of mercury (mmHg) down to about 10^{-5} mmHg. A sample of gas at this low pressure p is compressed by a known ratio in the gauge so as to give a much higher and easily measured pressure from which the original pressure may be calculated.

The tube T of the gauge (Fig. 9.2) is connected to the container of gas at pressure p. When the mercury reservoir is lowered sufficiently, the levels of mercury in all tubes fall below that of the junction J. Gas at the unknown pressure p then enters the bulb and closed capillary tube C_1. When the reservoir is raised the gas in the bulb is trapped and compressed into a much smaller volume in C_1. The reservoir is set so that the mercury level in C_2 is as shown at the zero level of the scale which extends downward from the top of C_1.

Then if h is the height of the column of gas trapped in C_1, this gas has pressure $(p+h\varrho g)$ and volume (ha), where ϱ is the density of the mercury, g is the acceleration of gravity, and a is

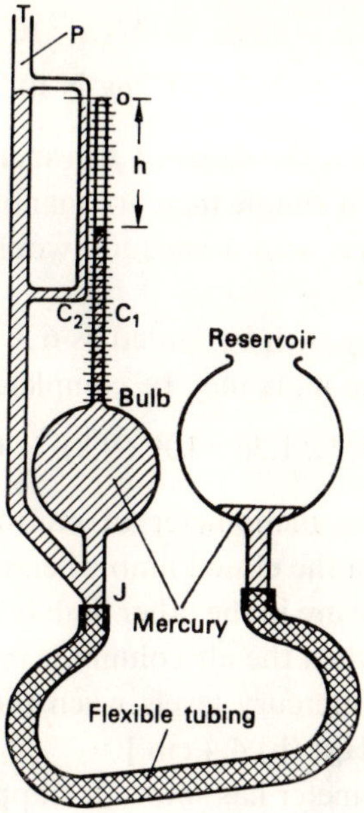

FIG. 9.2. The McLeod gauge (ex. 9.2).

the constant cross-section area of the bore of C_1. C_2 has the same value of a as C_1 so that surface tension does not affect the difference of levels h.

Now, from Boyle's law for constant temperature,

$$(p+h\varrho g)(ha) = pV, \tag{9.3}$$

where V was the volume of gas at pressure p as it became trapped at J when the reservoir was raised. Assuming $p \ll h\varrho g$, eqn. (9.3) simplifies to $(h\varrho g)(ha) = pV$. Then

$$p = h^2 a\varrho g/V. \tag{9.4}$$

Solution

From the data of the question, $h = 20\times10^{-3}$ m,

$$a = \pi\times(10^{-3})^2/4 \text{ m}^2, \quad \text{and} \quad V = 50\times(10^{-2})^3 \text{ m}^3.$$

$$\therefore \text{ from eqn. (9.4) } p = (20\times10^{-3})^2(\pi\times10^{-6}/4)\varrho g/(50\times10^{-6})$$
$$= 6.3\times10^{-6}\varrho g \text{ N m}^{-2}.$$

Thus 6.3×10^{-6} m is the value of p in metres of mercury. This difference of levels of a simple mercury manometer, registering the pressure in comparison with a vacuum, would be very difficult to measure.

The value of p may then be stated as 6.3×10^{-3} mmHg, or the calculation in absolute units may be completed:

$$p = 6.3\times10^{-6}\times1.36\times10^4\times9.81 = 0.84 \text{ N m}^{-2}.$$

9.3. A closed-tube manometer of constant bore has an air column 5.0 cm long in the closed limb when the mercury levels are equal and the gas pressure in the other limb is 200 mmHg. Calculate (a) the gas pressure when the air column is reduced to 2.0 cm, and (b) the difference of mercury levels when the gas pressure is 400 mmHg. [(a) 560 mmHg; (b) 4.4 cm.]

9.4. A faulty barometer has some air trapped over the mercury. The indicated height is 740 mm when the true height would be 760 mm, and is 760 mm when it would be 784 mm. Calculate the true height when 780 mm is indicated, assuming constant temperature. [810 mm.]

9.5. The McLeod gauge of ex. 9.2 has a scale on which the lowest pressure calibration mark is 1.5 mm from the end of the closed capillary. Calculate (a) the pressure which this point indicates, and (b) the compression ratio used in the gauge when measuring this pressure. [(a) 3.5×10^{-5} mmHg; (b) 4.2×10^4.]

9.6. The capillary tubes of a McLeod gauge have volumes of 1.0 mm³ per millimetre length, and the volume of gas as it becomes trapped in the bulb is 100 cm³. For an air column of length 10 mm in the closed capillary, calculate (a) the pressure of this air, and (b) the original pressure being measured. Ref.: mercury. [(a) 1.3×10^3 N m^{-2}; (b) 0.13 N m^{-3}.]

Constant Pressure or Volume

9.7. Worked Example

A mass of air, occupying a volume of 720 cm^3 at a temperature of 20°C is, heated to 100°C while the pressure is maintained constant. Calculate the volume at 100°C.

Introduction

Charles' law states that the volume of a constant mass of any gas at constant pressure increases by a constant fraction, 1/273, of the volume at 0°C per deg C rise of temperature. Alternatively, it may be said that the temperature coefficient of volumetric expansion at constant pressure is the same for all gases when referred to the same temperature. Also the law of pressure variation at constant volume is of the same form, with the same coefficient 1/273.

The above statements are only approximate for real gases, but may be used in many cases of substances above their boiling points. An ideal gas would follow the laws of Boyle and Charles exactly at all temperatures.

It follows from the statement of Charles' law that the expansion

$$\delta V = (V_0/273)\,\delta\theta, \tag{9.5}$$

when V_0 = volume at 0°C, $\delta\theta$ = rise of temperature in deg C at constant pressure, and 1/273 per deg C is the temperature coefficient of volumetric expansion at constant pressure [compare with eqn. (7.8)]. Then the volume at any temperature θ is

$$V = V_0 + \delta V = V_0[(1+(\theta/273)]. \tag{9.6}$$

This equation is simplified by changing the zero level of the temperature scale to the "absolute zero" where V would reach zero if the law still held. Thus Fig. 9.3 shows temperature θ on the Celsius scale and temperature T on the Kelvin scale which has the absolute zero and the same size division or degree as the Celsius scale. The abbreviation for Kelvin is K, no symbol being required for "degree" as in °C or deg C.

Then, since $T = \theta + 273$, eqn. (9.6) becomes

$$V = V_0[1+(T-273)/273] = V_0T/273.$$

$$\therefore \; V/T = V_0/273 = \text{a constant.} \tag{9.7}$$

Thus for a constant mass of an ideal gas at constant pressure, Fig. 9.3 and eqn. (9.7) show that the volume is proportional to the absolute temperature.

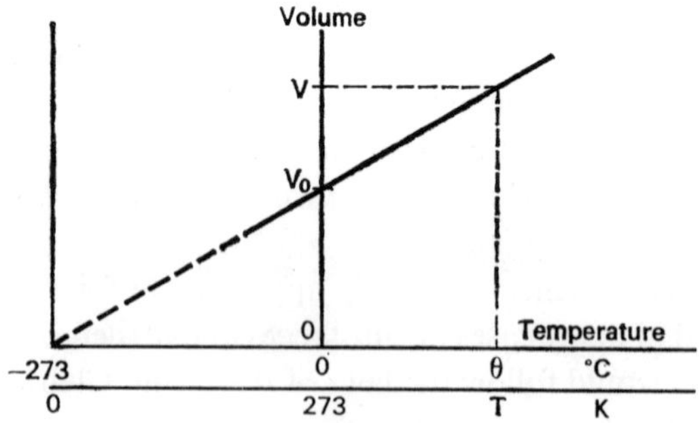

FIG. 9.3. Temperature–volume curve.

Similarly, taking the law of pressure variation for a constant mass of ideal gas at constant volume to have the temperature coefficient 1/273 per deg C, it follows that

$$p/T = p_0/273 = \text{a constant} \tag{9.8}$$

for a constant mass of ideal gas at constant volume.

Solution

In the first condition $T_1 = 20+273 = 293$ K, and $V_1 = 720$ cm^3.

In the second condition $T_2 = 100+273 = 373$ K.

Then, from eqn. (9.7), $V_1/T_1 = V_2/T_2$.

Therefore the required volume $V_2 = (V_1/T_1)T_2 = (720/293)373 = 920$ cm^3.

9.8. The volume of a certain mass of CO_2 is 0.20 m^3 at a temperature of 50°C. Calculate the volume at the same pressure if the temperature on the Celsius scale is doubled. [0.23 m^3.]

9.9. A car tyre is inflated at a temperature of 15°C to a gauge pressure of 2.0×10^5 N m^{-2}. Calculate the gauge pressure when the temperature rises to 35°C, with constant atmospheric pressure of 1.0×10^5 N m^{-2}. Assume that the volume of the tyre remains constant. [2.2×10^5 N m^{-2}.]

9.10. A pressure gauge connected to a sealed container of gas at 60°C indicates a gauge pressure of 200 mmHg against an atmosphere of standard pressure. Calculate the temperature when the indication of the gauge is 520 mmHg. Ref.: standard pressure. [170°C.]

Single Gas Equation

9.11. Worked Example

The volume of a certain mass of gas is 2.50 m^3 at a temperature of 0°C and a pressure of 740 mmHg. Calculate the volume when the temperature is changed to 100°C and the pressure to 780 mmHg.

Introduction

The laws of Boyle and Charles may be stated by a single equation relating the volume V, the absolute pressure p, and the absolute temperature T, obtained as follows. From Boyle's law,

$V \propto 1/p$, [see eqn. (9.1)], and from Charles' law, $V \propto T$ [see eqn. (9.7)]. Then, by combination of these two proportionalities,

$$V \propto T/p \tag{9.9}$$

for a constant mass of ideal gas. This result may be checked by putting T = constant for Boyle's law and p = constant for Charles' law.

It follows from (9.9) that pV/T is constant for a constant mass. Thus for two conditions, 1 and 2, for the same mass of gas,

$$p_1V_1/T_1 = p_2V_2/T_2. \tag{9.10}$$

It should be noted that eqn. (9.10) cannot be obtained by manipulating the equations $pV = c_1$ and $V/T = c_2$ as if they were simultaneous equations stating the individual laws with constants c_1 and c_2. These equations are not true simultaneously, since T is constant for the first but variable in the second, and p is variable in the first but constant for the second. Thus, for example,

$$(pV)(V/T) \neq c_1c_2 \quad \text{and} \quad (pV)+(V/T) \neq c_1+c_2.$$

Solution

In the particular case,

$$p_1 = 740 \text{ mmHg}, \quad V_1 = 2.50 \text{ m}^3, \quad T_1 = 273 \text{ K},$$
$$p_2 = 780 \text{ mmHg}, \quad V_2 \text{ is unknown}, \quad T_2 = 373 \text{ K}.$$

Then, from eqn. (9.10), $V_2 = (p_1V_1/T_1)\times(T_2/p_2)$
$= (740\times2.50/273)\times(373/780) = 3.24 \text{ m}^3.$

Alternatively, the changes of pressure and temperature may be considered separately, one after the other, with an intermediate volume V_x.

Thus for a change of pressure at constant temperature T_1, Boyle's law gives $V_x = V_1p_1/p_2 = 2.50\times740/780 = 2.37 \text{ m}^3$.

Then for a subsequent change of temperature at constant pressure p_2, Charles' law gives the final volume

$$V_2 = V_x T_2/T_1 = 2.37\times 373/273 = 3.24 \text{ m}^3.$$

9.12. The volume of a fixed mass of gas at a temperature of 10°C is 250 cm³. Calculate the volume at 45°C with double the original pressure. [140 cm³.]

9.13. Convert the following volumes to s.t.p.: (a) 3.0 l. at 15°C and 1500 mmHg, and (b) 5.0 m³ at −20°C and 1.5×10^5 N m^{-2}. Ref.: s.t.p. [(a) 5.6 l.; (b) 8.0 m³.]

9.14. Some air at 60°C and 760 mmHg is heated to 250°C. What pressure is then needed to reduce the volume to two-thirds of the original? [1800 mmHg.]

9.15. A mercury manometer has one end closed and the other connected to a vessel containing gas. When the temperature is 30°C, the air column in the closed tube is 5.0 cm long and the gauge pressure is −15 mmHg. When the vessel is opened to the atmosphere at standard pressure and the temperature is 10°C, the gauge pressure is +10 mmHg. Calculate the original gas pressure, neglecting any change in the density of mercury. [590 mmHg.]

Change of Density

9.16. Worked Example

Calculate the ratio of densities of air under the following conditions of temperature and pressure: (a) 10°C, 780 mmHg, (b) 150°C, 2000 mmHg.

Introduction

As in eqn. (9.10), $p_1V_1/T_1 = p_2V_2/T_2$ for two conditions of a constant mass m of gas. Now dividing both sides of this equation by m gives

$$(p_1V_1)/(mT_1) = (p_2V_2)/(mT_2).$$

But $V_1/m = 1/\varrho_1$ and $V_2/m = 1/\varrho_2$, where ϱ = density.

$$\therefore\ p_1/\varrho_1 T_1 = p_2/\varrho_2 T_2. \tag{9.11}$$

This applies whether the mass is constant or not, since on either side of the equation $1/\varrho$ is the volume per unit mass.

Solution

From eqn. (9.11) $\varrho_1/\varrho_2 = (p_1/p_2)\times(T_2/T_1)$
$= (780/2000)\times(150+273)/(10+273) = 0.58.$

9.17. Calculate the density of air at a temperature of 80°C and a gauge pressure of 150 mmHg. Ref.: air. [1.2 kg m^{-3}.]

9.18. The dimensions of a rectangular room are $7.0\times10.0\times 3.0$ m. Calculate the mass of air in the room at a temperature of 24° C and a pressure of 730 mmHg. Ref.: air. [240 kg.]

9.19. Two vessels, each containing 2 l. of air at s.t.p., are connected by a narrow tube. Calculate the mass of air displaced from one vessel to the other when one is heated to 100°C while the other is maintained at 0°C. Ref.: air [0.4 g.]

Ideal Gas Scale of Temperature

9.20. Worked Example

A constant volume helium thermometer indicates a gauge pressure of −300 mmHg relative to standard atmospheric pressure when the temperature is 0°C. Calculate the ideal gas temperature when the gauge pressure is −350 mmHg.

Introduction

The indications of thermometers employing different substances or different properties of substances do not agree exactly when each thermometer has a uniform or linear scale (see ex. 7.1). This applies also to gas thermometers in which expansion or increase of pressure is used, since real gases differ from each other and only approximately follow the ideal gas laws.

However the indications of low pressure gas thermometers tend toward ideal agreement as the pressures are reduced. Moreover, the kinetic theory of gases (see ex. 15.14) shows that the pressure of an ideal gas at constant volume and ideally low pressure is directly related to the energy concept of temperature.

The ideal gas scale can then be described as that which would make the expression pV/T exactly constant for a real gas at ideally low pressure.

The standard type of thermometer for this scale is a "constant volume" thermometer in which variation of gas pressure is measured at almost constant volume. Correction is made for the slight expansion of the gas container and for the fact that the pressure is not ideally low.

The fixed points of the scale are 0 K when $p = 0$, and 273.16 K when water is at its triple point, at which pure ice is in equilibrium with pure water and pure water vapour with no air. Then if p_t is the corrected pressure at the triple point and p is the value at temperature T for constant volume,

$$p/T = p_t/273.16. \tag{9.12}$$

On the Celsius scale with the ice point at 0.00°C and the steam point at 100.0°C the triple point of water is at 0.01°C. The difference between the ice point and the triple point is due to the exclusion of air at the triple point.

Figure 9.4 shows both Celsius and Kelvin scales, with

$$T = \theta + 273.15 \quad \text{and} \quad p \propto T. \tag{9.13}$$

Solution

At the ice point, $p_0 = 760 - 300 = 460$ mmHg, where 760 mmHg is the standard atmospheric pressure. At the unknown temperature, $p = 760 - 350 = 410$ mmHg.

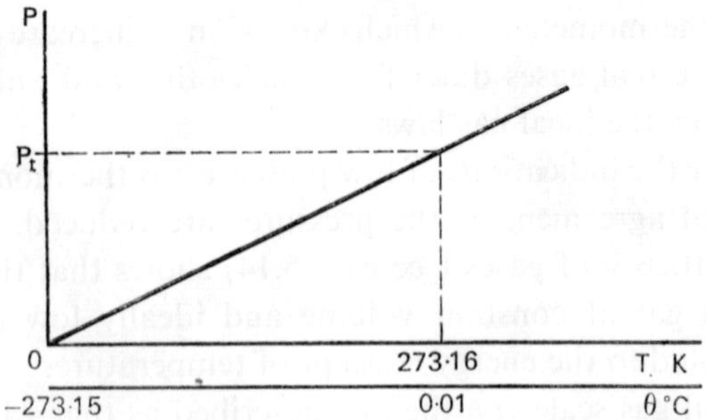

FIG. 9.4. Celsius and Kelvin temperature scale.

Then, assuming the gas behaviour to be ideal, eqn. (9.13) gives

$$T = 273.15(410/460) = 243 \text{ K}$$
$$= 243 - 273°\text{C} = -30°\text{C}.$$

9.21. The pressure of the gas of a constant volume gas thermometer is found to be 150 mmHg at the steam point. Calculate the ideal gas temperature when the pressure is 200 mmHg, assuming the thermometer to contain an ideal gas. [224°C.]

9.22. Calculate the temperature in degrees Celsius on the linear scale of a constant volume gas thermometer when the pressure is 800 mmHg. Assume that the pressures at the ice and steam points have been determined as 700 mmHg and 956 mmHg respectively. [39°C.]

CHAPTER 10

QUANTITY OF HEAT

Specific Heat

10.1. Worked Example

An aluminium can of mass 46 g contains 30 g of water at a temperature of 10°C. When 150 g of lead at 80°C are placed in the water, the temperature of the water rises to 17°C. Calculate (a) the water equivalent of the can, (b) the thermal capacity and specific heat of the lead, and (c) the final temperature which would be reached if the mass of lead were doubled. Ref.: aluminium and water.

Introduction

It was seen in ex. 7.1 that temperature is a measure of the level of thermal activity of the molecules of a body such that a difference of temperature of two bodies leads to a flow of heat if they are put in contact. Individual collisions between the molecules may convey energy either way locally at different times, but, by and large, the more active molecules at the higher temperature lose energy to those at the lower temperature by increasing their activity.

If a certain quantity of heat energy is acquired by a body, the effect on its temperature must depend upon the total mass of the molecules sharing in it and upon the extent to which some of the energy is held as potential energy in the attractions within the body, or is used in work during expansion against external forces. When there is a change of state, as in ex. 10.9, such increases of potential energy and external work require heat even without rise of temperature, but if the body is solid or liquid and remains in that state, the quantity Q of heat required to raise the temperature is given by

$$Q = mc\,\Delta\theta, \tag{10.1}$$

where m is the mass of the body, $\Delta\theta$ is the temperature rise, and c is the "specific heat" which depends upon the nature of the material and is approximately a constant for a limited range of temperature.

The thermal capacity of a body is the heat gained by the body per unit rise of temperature. So, from eqn. (10.1),

$$\text{thermal capacity} = Q/\Delta\theta = mc. \tag{10.2}$$

The water equivalent of a body of mass m is the mass m_w of water which has the same thermal capacity. So, from eqn. (10.2), $mc = m_w c_w$, where c_w is the specific heat of water.

$$\therefore \text{ water equivalent} = mc/c_w. \tag{10.3}$$

Equations (10.1), (10.2), and (10.3) apply in any consistent set of units, including M.K.S. units, in which Q is in joules, m is in kilograms, $\Delta\theta$ in degrees Celsius (or Kelvin), and c is in J kg^{-1} deg C^{-1} (or J kg^{-1} K^{-1}). In M.K.S. units the specific heat of water c_w has a mean value between 0°C and 100°C of 4.19×10^3 J kg^{-1} deg C^{-1}, in agreement with $c_w = 1$ cal g^{-1} deg C^{-1}, and 1 cal = 4.19 J.

Solution

(a) Using M.K.S. units, $m = 46\times10^{-3}$ kg and $c = 910$ J kg^{-1} deg C^{-1} for aluminium compared with $c_w = 4190$ J kg^{-1} deg C^{-1} for water. Then eqn. (10.3) gives

$$m_w = 46\times10^{-3}\times910/4190 = 1.0\times10^{-2}\text{ kg} = 10\text{ g}. \tag{a}$$

(b) When the lead is placed in the water there is a transfer of heat from the lead to the water and aluminium, until all reach the same temperature of 17°C. Assuming no wastage of heat or gain from outside, the heat lost by the lead equals the heat gained by the water and aluminium. But the water and aluminium are equivalent to a mass of water of $30+10 = 40$ g. Then, using eqn. (10.1) with M.K.S. units again, heat gained $= (40\times10^{-3})\times4190\times(17-10) = 1170$ J. Since this is lost by the lead, eqn. (10.2) gives thermal capacity of lead

$$Q/\Delta\theta = 1170/(80-17) = 18.6 = 19 \text{ J deg C}^{-1} \quad \text{(2 sig. fig.).} \quad \text{(b)}$$

Also from eqn. (10.2),

$$c = (Q/\Delta\theta)/m = 18.6/0.150 = 124$$
$$= 120 \text{ J kg}^{-1} \text{ deg C}^{-1} \quad \text{(2 sig. fig.).} \qquad \text{(b)}$$

(c) When the mass of lead is doubled, again assuming no wastage or net gain of heat,

$$(mc\,\Delta\theta) \text{ for lead} = (mc\,\Delta\theta) \text{ for total equivalent mass of water.}$$

$$\therefore\ 0.300\times124\times(80-\theta) = 0.040\times4190\times(\theta-10),$$

where θ = final temperature.

$$\therefore\ 2980-37.2\theta = 167\theta-1670.$$

$$\therefore\ \theta = (2980+1670)/(167+37.2) = 4650/204.2 = 23°\text{C}. \quad \text{(c)}$$

10.2. Worked Example

A mass of 50 g of copper at 100°C is placed in a mass of water of 60 g at 15°C, contained in a copper can of mass 160 g. The final temperature is 20°C. Calculate (a) the specific heat of copper assuming no wastage of heat, and (b) the time required to heat the can and its contents from 20 to 100°C by means of a heater of power 100 W, assuming an efficiency of 75%.

Introduction

When the joule is the unit of heat energy, the watt is the corresponding unit of rate of flow of heat or heat power. Then the heat Q provided in a time t by a heater of constant power P is given by

$$Q = Pt, \tag{10.4}$$

where Q is in joules (J), P is in watts (W), and t in seconds (s). If some heat is wasted, the heat used is given by

$$\eta Pt \tag{10.5}$$

where η is the efficiency of use expressed in decimal or fractional form, e.g. $\eta = 0.75$ or 3/4 for an efficiency of 75%.

Solution

(a) If no heat is wasted, the heat lost by the copper equals that gained by the can and water. Thus using eqn. (10.1) for each of these three items,

$$0.050c(100-20) = [0.160c(20-15)]+[0.060\times 4190\times(20-15)].$$

$$\therefore\ 4c = 0.8c+1257.$$

$$\therefore\ c = 1257/3.2 = 390 \text{ J kg}^{-1} \text{ deg C}^{-1}. \tag{a}$$

(b) For a rise of 80 deg C for the water and all the copper, eqn. (10.1) gives

$$Q = (0.060\times 4190\times 80)+[(0.160+0.050)\times 390\times 80]$$
$$= (2.01\times 10^4)+(0.66\times 10^4) = 2.67\times 10^4 \text{ J}.$$

Then, from eqn. (10.5),

$$2.67\times 10^4 = \eta Pt = 0.75\times 100\times t.$$

$$\therefore \text{ time required, } t = 2.67\times 10^4/75 = 360 \text{ s} = 6.0 \text{ min.} \tag{b}$$

Note that there is often no advantage in considering the water equivalent of a body in M.K.S. units. Thus if eqn. (10.3) is used to give the water equivalent of all the copper,

$$m_w = mc/c_w = (0.050+0.160)\times 390/4190.$$

Then the number 4190 will cancel out again when eqn. (10.1) is used.

10.3. A quantity of heat of 1170 J is required to raise the temperature of a mass of copper of 150 g from 50 to 70°C. Calculate the thermal capacity, the specific heat, and the water equivalent of the copper. [59 J deg C^{-1}; 390 J kg^{-1} deg C^{-1}; 14 g.]

10.4. Calculate the heat required to raise the temperature of 370 g of water from 15° to 100°C in a can of mass 150 g and specific heat 900 J kg^{-1} deg C^{-1}. [1.4×10^5 J.]

10.5. A mass of 270 g of iron at a temperature of 90°C is placed in a vessel of water having a total water equivalent of 60 g. The temperature of the water rises from 7° to 35°C. Calculate the thermal capacity and specific heat of the iron. Ref.: water. [130 J deg C^{-1}; 470 J kg^{-1} deg C^{-1}.]

10.6. It is required to cool 25,000 l. of oil per hour from 60° to 25°C. Calculate the rate of flow of water required to cool the oil when the initial temperature of the water is 15°C. Assume relative density of oil = 0.90 and specific heat of oil = 2.0×10^3 J kg^{-1} deg C^{-1}. Ref.: water. [3.8×10^4 l. h^{-1}.]

10.7. A mass of 200 g of water at 90°C is added to 300 g of water at 15°C in an aluminium can of mass 100 g. Calculate final temperature. Ref.: aluminium and water. [44°C.]

10.8. In an experiment to determine the specific heat of water by continuous flow, a rise of temperature of 2.40 deg C occurred when the rate of flow was 57 cm^3 min^{-1}, and the heater supply was 2.0 A at 4.0 V. The same temperature rise was obtained for a rate of flow of 78 cm^3 min^{-1} and a heater supply of 2.4 A at 4.8 V. Assuming the same rate of loss of heat in each case, calculate the specific heat of water. [4.2×10^3 J kg^{-1} deg C^{-1}.]

Latent Heat

10.9. Worked Example

A mass of 25 g of steam at 100°C is condensed on a mixture of 100 g of ice and 75 g of water at 0°C. Calculate the rise of temperature of the mixture.

Introduction

When a liquid turns to vapour, the molecules have to break free of forces attracting them back as they leave the liquid, (see also surface tension, ex. 8.25), and to force back the atmosphere to make room for the larger volume of vapour. Since many molecules in the liquid have insufficient energy to escape, the mean thermal energy of molecules in the liquid falls as the faster molecules leave. So heat energy must be supplied to cause a liquid to evaporate or boil at constant temperature.

Similarly, when a solid melts, the molecules or atoms must break free from the solid structure, and heat is required for this to occur at constant temperature. Conversely, heat is released when a vapour condenses at constant temperature, or a liquid freezes at constant temperature. When such a change of state is accompanied by a change of temperature, the two changes lead to separate terms in the calculation of the total heat gained or lost.

The heat required to change the state of unit mass of substance at constant temperature is called the latent heat of that change. Thus the latent heat of fusion (melting) of ice is 3.35×10^5 J kg^{-1} at 0°C, and the latent heat of vaporization of water is 2.25×10^6 J kg^{-1} at 100°C.

Solution

When the steam condenses on the ice and water, the heat lost by the steam is due to both its change of state to water and the cooling of this water (condensate). Therefore the heat lost by 0.025 kg of steam $= (0.025\times 2.25\times 10^6)+(0.025+4190\times(100-\theta))$

$= 66700-1050\theta$ (3 sig. fig.),

where the specific heat of water is taken to be 4190 J kg^{-1} deg C^{-1}, and θ is the final temperature.

Heat is gained by the ice in the change to water, and further heat is gained by this water and the original water during the subsequent rise of temperature. Therefore the heat gained by the ice and water

$= (0.100\times 3.35\times 10^5)+(0.100\times 4190\theta)\times(0.075\times 4190\theta)$

$= 33500+733\theta$ (3 sig. fig.).

Then assuming no external loss or gain,

$$66700-105 = 33500+733\theta.$$

$$\therefore\ \theta = 33200/838 = 40°\text{C} \quad \text{(2 sig. fig.)}.$$

Since the initial temperature was 0°C, the rise is 40 deg C.

10.10. Worked Example

A well-lagged iron can of mass 150 g contains 150 g of water at 20°C. A block of aluminium of mass 300 g is heated to 350°C and then placed in the water. Calculate the mass of steam produced at normal atmospheric pressure. Ref.: latent and specific heats.

Solution

Four quantities of heat are involved in the exchanges:

(a) the heat given up by the aluminium;
(b) the heat taken by the can;
(c) the heat taken by the water rising to the boiling point;
(d) the heat taken by the water turning to steam at the boiling point.

Then, assuming no external loss or gain, (a) = (b)+(c)+(d). Here (a), (b), and (c) are calculated by using eqn. (10.1) with values of c given in the reference table.

Thus heat (a) given up by aluminium cooling from 350°C to boiling point of water, 100°C, is

$$mc\,\Delta\theta = 0.300\times 910\times(350-100)$$
$$= 68.2\times 10^3 \text{ J.} \qquad \text{(a)}$$

Also heat (b) taken by can $= 0.150\times 470\times(100-20)$

$$= 5.64\times 10^3 \text{ J.} \qquad \text{(b)}$$

Similarly, heat (c) taken to raise temperature of water to 100°C

$$= 0.150\times 4190\times(100-20)$$
$$= 50.3\times 10^3 \text{ J.} \qquad \text{(c)}$$

Lastly, heat (d) to produce steam from water at 100°C

$$= m\times(\text{latent heat of vaporization}) = m\times 2.25\times 10^6 \text{ J.} \qquad \text{(d)}$$

But (d) = (a)−(b)−(c).

$$\therefore\ m\times 2.25\times 10^6 = (68.2-5.64-50.3)\times 10^3 \text{ J.}$$

$$\therefore\ m = (68.2-55.94)\times 10^{-3}/2.25 \text{ kg} = 5.5 \text{ g.}$$

10.11. Calculate the rate at which ice melts when heated by a 500 W heater. Assume 100% efficiency. Ref.: latent heat. [1.5 g s^{-1}.]

10.12. Calculate the rise of temperature produced by a mass of 4.0 g of steam at 100°C condensing in a can of water equivalent 10 g containing 120 g of water at 10°C. Ref.: latent heat of steam. [19 deg C.]

10.13. A mass of 10 g of ice at a temperature of −5°C is mixed with 160 g of water at 10°C contained in a can of mass 150 g. The specific heat of ice = 0.50 cal g^{-1} deg C^{-1}, and the specific heat of the can material = 0.20 cal g^{-1} deg C^{-1}. Ref.: latent heat. Calculate the final temperature. [5.4°C.]

10.14. A piece of glass of mass 25 g and temperature 15°C is placed in a steam calorimeter. The steam temperature is 100°C and the mass of condensate is 0.63 g. Calculate the specific heat of the glass. Ref.: steam. [670 J kg^{-1} deg C^{-1}.]

10.15. Calculate the final temperature when 50 g of ice at 0°C is added to 30 g of water at 40°C in a can of mass 100 g and specific heat 400 J kg^{-1} deg C^{-1}. Ref.: latent heat of ice. [0°C.]

10.16. An electric heater in a beaker of water makes a total mass of 300 g at 15°C. The total water equivalent is 290 g. Electric energy is supplied at a constant rate, and steam can escape freely, although loss of heat in other ways is negligible. After the first 10 min the water boils. After a further 15 min the total weight is found to be 233 g. Calculate the latent heat of steam. [2.3×10^6 J kg^{-1}.]

10.17. An immersion heater is used to keep a liquid boiling in an open can. When the heater power is 25 W, the mass of liquid decreases by 250 mg in a time of 30 s, and when 50 W by 590 mg in 30 s. Calculate the latent heat of the liquid assuming a constant rate of loss of heat from the sides and base of the can. [2.2×10^6 J kg^{-1}.]

Heat and Work

10.18. Worked Example

A vehicle of mass 1500 kg accelerates on a horizontal road at a rate of 1.5 m s^{-2} for a distance of 40 m. (a) Calculate the heat loss during the acceleration, assuming an efficiency of 20% and no other loss but heat, and (b) calculate also the heat developed in stopping the vehicle from a speed of 60 km h^{-1} using friction brakes.

Introduction

When considering the conversion of energy from one form to another, or the transfer of energy in work or flow of heat, it is most convenient to have all quantities of energy in the same units, namely joules in the M.K.S. system. However, heat is frequently measured in calories, one calorie being the heat required to raise the temperature of one gram of water through one degree Celsius. The conversion ratio, known as Joule's equivalent, is 4.19 J cal^{-1}, i.e. 1 cal = 4.19 J.

Solution

(a) The resultant force causing acceleration is given by

$$(\text{mass})\times(\text{acceleration}) = 1500\times 1.5 = 2250 \text{ N}.$$

Then the useful work done in acceleration is given by

$$(\text{resultant force})\times(\text{distance}) = 2250\times 40 = 9.0\times 10^4 \text{ J}.$$

Since this is 20% or one-fifth of the total energy used by the engine, the losses are 4 times greater, i.e. 80% or four-fifths of the total energy.

$$\therefore \text{ heat loss} = 4\times 9.0\times 10^4 = 3.6\times 10^5 \text{ J}. \qquad \text{(a)}$$

This may be stated as $3.6\times 10^5/4.19 = 8.6\times 10^4$ cal. (a)

(b) The heat developed in stopping the vehicle is due to the conversion of kinetic energy to heat in friction brakes.

Now kinetic energy $= \frac{1}{2}mv^2$ joules, where m = mass = 1500 kg and velocity $v = (60\times 10^3)/(3.6\times 10^3)$ m s^{-1} = 16.7 m s^{-1}.

$$\therefore \text{ heat developed} = \tfrac{1}{2}\times 1500\times(16.7)^2 = 2.1\times 10^5 \text{ J}. \qquad \text{(b)}$$

This may be stated as $2.1\times 10^5/4.19 = 5.0\times 10^4$ cal. (b)

10.19. Calculate the rise of temperature of the water of a river due to a drop of 85 m at a water fall. Assume the velocity of the water to be the same when arriving and leaving. Ref.: water. [0.2 deg C.]

10.20. A vehicle of mass 4000 kg is brought to rest from a speed of 54 km h^{-1} in a distance of 60 m by a constant friction force applied by the brakes. Calculate (a) the initial rate, and (b) the average rate at which heat is produced in the brakes. Assume forces of resistance other than those of the brakes to be neglig ble. [(a) 112 kW; (b) 56 kW.]

10.21. A crate of mass 50 kg is dragged over a distance of 17 m along a horizontal floor. The coefficient of friction with the floor is 0.30. Calculate the heat produced. [2.5×10^3 J.]

10.22. A quantity of 1.0 l. of water in an aluminium can of mass 200 g is stirred by a paddle driven by a constant torque of 2.0 Nm. Calculate the number of revolutions required to raise the temperature by 10 deg C, assuming an efficiency of 40%. Ref.: specific heat. [8700 rev.]

10.23. A motor is coupled to a hydraulic brake consisting of paddles rotating in a vessel of oil. The mass of oil is 0.50 kg and its specific heat is 2000 J kg^{-1} deg C^{-1}. The container has a water equivalent of 0.30 kg, and the paddles a water equivalent of 0.20 kg. The temperature rises at a rate of 5.0 deg C min^{-1} when the motor runs at a certain constant rate. Calculate the output power of the motor, assuming constant resistance to motion due to the brake and an efficiency of 85% to allow for heat loss from the brake. Ref.: water. [0.30 kW.]

Calorific Values

10.24. Worked Example

Calculate how long 1.0 tonne of coal would last when used to supply heat to an engine of 20 kW with an overall efficiency of 15%. Assume the calorific value of the coal to be 3.0×10^7 J kg^{-1}.

Introduction

The calorific value of a fuel is the amount of heat produced by the complete combustion of unit quantity of the fuel when no unburnt fuel escapes in smoke or ash. Then the input heat energy to the engine due to complete combustion is given by the calorific value multiplied by the mass of fuel used. In practice, of course, some fuel is wasted and a large proportion of the heat produced is wasted.

Solution

The maximum energy obtainable from 1.0 tonne (= 1000 kg) of the coal is $3.0\times10^7\times1000 = 3.0\times10^{10}$ J, of which only 15% appears as useful output energy.

$$\therefore \text{ useful output} = 0.15\times3.0\times10^{10}$$
$$= 0.45\times10^{10} \text{ J per tonne.}$$

When the engine works for a time t in seconds at a rate of 20 kW, the useful work done $= t\times20\times10^3$ J.

Therefore 1.0 tonne of coal would last for a time t where

$$t\times20\times10^3 = 0.45\times10^{10}. \qquad \therefore t = 2.25\times10^5 \text{ s.}$$

$$\therefore \text{ time} = 2.25\times10^5/3.6\times10^3 = 63 \text{ h} \quad \text{(2 sig. fig.)}$$

10.25. A gas burner consumes gas at a rate of 250 l h^{-1}. Calculate the rate at which heat is produced assuming complete combustion of the gas and a calorific value of 2.0×10^7 J m^{-3}. [1.4 kW.]

10.26. Calculate the output power of a steam engine of 30% efficiency when coal of calorific value 8.0×10^6 cal kg^{-1} is used at a rate of 100 kg h^{-1}. [280 kW.]

10.27. Calculate the efficiency of a petrol engine which uses fuel at a rate of 11 l. h^{-1} when working at a power of 30 kW. Assume the calorific value of the petrol to be 4.5×10^7 J kg^{-1}, and the relative density to be 0.80. [27%.]

10.28. Calculate the rate of consumption of coal for the production of steam at a rate of 2.0 kg h^{-1} at normal atmospheric pressure from water at a temperature of 10°C. Assume the calorific value of the coal to be 3.0×10^7 J kg^{-1} and the efficiency to be 60%. Ref.: specific and latent heat. [0.29 kg h^{-1}.]

10.29. A specimen of coal of mass 1.04 g is burned completely in oxygen in a bomb calorimeter which retains all products of combustion. The calorimeter has a water equivalent of 486 g and contains 2.31 l. of water additionally. The temperature rise due to the combustion is 2.66°C. Calculate the calorific value of the specimen. Ref.: water. [3.0×10^7 J kg^{-1}.]

CHAPTER 11

CENTRIPETAL ACCELERATION

Centripetal Force and Acceleration

11.1. Worked Example

A man of mass 75 kg is seated in a car which turns a corner of a road at a constant speed of 48 km h^{-1}. If the man moves in a curve of radius 50 m, calculate (a) his centripetal acceleration, and (b) the resultant force which must act on him.

Introduction

In accordance with Newton's first law of motion, a body travelling in a straight line will go straight on if no force is available to turn its direction of motion. If the body is to move along a curve at constant speed, a resultant force is required to act on the body at every point of the curve in such a direction as to turn the velocity continuously without increasing or decreasing the speed. So the force must be centripetal, i.e. directed always towards the centre of curvature of the path, as shown by F_1, F_2, etc., in Fig. 11.1 which shows an S-shaped path. At any instant this direction is perpendicular to the tangent T_1 or T_2, etc., along which the body would move if the resultant force suddenly became zero.

In accordance with Newton's second law, the effect of a resultant centripetal force F_c on a body of mass m must be the production of a centripetal acceleration

$$a_c = F_c/m. \qquad (11.1)$$

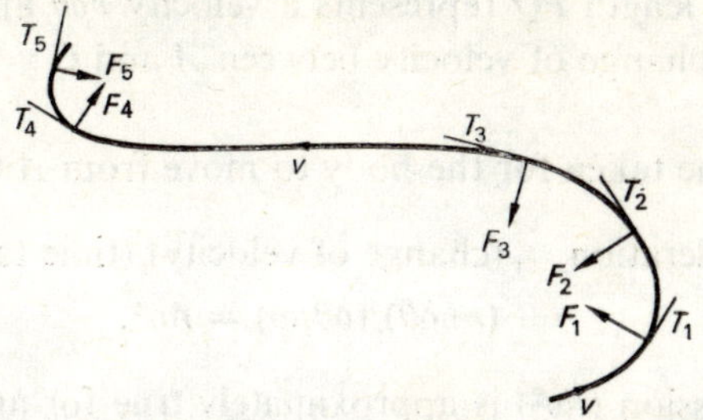

FIG. 11.1. Centripetal force and direction of motion.

Thus the continuous turning of the velocity amounts to a continuous addition of velocity perpendicular to the tangent to the path at each instant, and so amounts to a centripetal acceleration at constant speed.

The centripetal acceleration may be found in terms of the speed and radius of the path by first considering the path to be a number of small steps such as *ABCDE* in Fig. 11.2. *AB* and *CD* are tangents in the directions of the velocities v_A at *A* and v_C at *C*; *BC* and *DE* are in the directions of the radii at *C* and *E*. The triangle of velocities shown in Fig. 11.2 is isosceles, since v_A and v_C have the same speed $v = r\omega$, where r = radius = $OA = OC$, and ω = angular velocity of body around curve *ACE*. If $\delta\theta$ is very small, in radians, then the third side of the velocity triangle which approximately

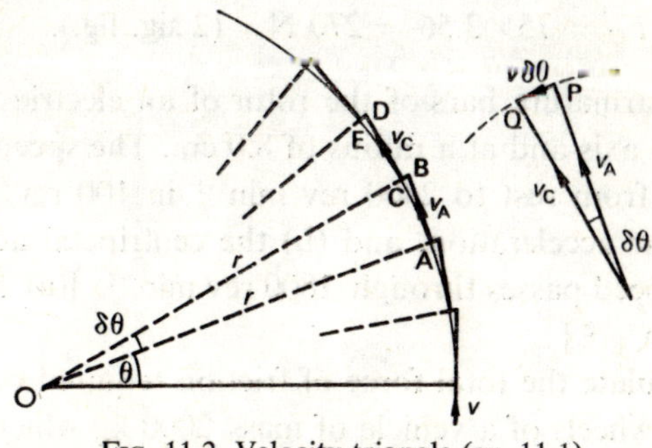

FIG. 11.2. Velocity triangle (ex. 11.1).

equals the arc length PQ represents a velocity $v\delta\theta$ approximately. Therefore the change of velocity between A and $C \simeq v\delta\theta = r\omega\delta\theta$, since $v = r\omega$.

But the time taken for the body to move from A to $C = \delta\theta/\omega$.

$$\therefore \text{ acceleration} = (\text{change of velocity})/(\text{time taken})$$
$$\simeq (r\omega\delta\theta)/(\delta\theta/\omega) = r\omega^2.$$

This expression ($r\omega^2$) is approximately true for any very small tooth of the above motion. If each tooth is made smaller the approximation improves until in the limit it would be exactly true for infinitely small teeth. Then, since circular motion is the same as an infinite number of these infinitesimal steps, the expression is exactly true for perfectly circular motion.

$$\therefore \text{ centripetal acceleration } a_c = r\omega^2 = v^2/r. \qquad (11.2)$$

Then, with eqn. (11.1), centripetal force $F_C = mr\omega^2 = mv^2/r$. (11.3)

Solution

(a) For a speed of $48/3.6$ m s^{-1} around a curve of radius 50 m, eqn. (11.2) gives

$$a_c = (48/3.6)^2/50 = 3.56 = 3.6 \text{ m s}^{-2} \quad (2 \text{ sig. fig.}). \qquad (a)$$

(b) For a mass of 75 kg, eqn. (11.1) or (11.3) then gives

$$F_C = 75 \times 3.56 = 270 \text{ N} \quad (2 \text{ sig. fig.}). \qquad (b)$$

11.2. The armature bars of the rotor of an electric motor are parallel to the axis and at a radius of 8.0 cm. The speed rises at a constant rate from rest to 2400 rev min^{-1} in 100 rev. Calculate (a) the angular acceleration, and (b) the centripetal acceleration as the rotor speed passes through 1800 rev min^{-1}. [(a) 50 rad s^{-2}; (b) 2.8×10^3 m s^{-2}.]

11.3. Calculate the total force of friction required between the road and the wheels of a vehicle of mass 2000 kg which rounds a bend of radius 50 m at a speed of 25 km h^{-1} on a level surface. [1.9×10^3 N.]

11.4. A wheel requires the addition of a mass of 20 g on the rim, of diameter 50 cm, for dynamic balance. Calculate the out-of-balance force on the bearings at a speed of 1000 rev min^{-1} before the wheel is balanced. [55 N.]

11.5. Calculate the acceleration of a body travelling at a constant speed around a circular arc when in a time of 1.0 ms the distance traversed is 3.0 cm and the change of direction is 30°. [1.6×10^4 m s^{-2}.]

11.6. A horizontal beam of length 4.0 m carries experimental apparatus at one end and rotates about a vertical axis at the other end. Calculate the rate of rotation to test the apparatus under horizontal acceleration five times greater than gravity. [33 rev min^{-1}.]

11.7. Calculate the minimum radius of a curve which a vehicle can follow at a speed of 50 km h^{-1} without skidding on a level surface when the coefficient of friction is 0.50. [39 m.]

Gravitation and Circular Orbits

11.8. Worked Example

Calculate the minimum vertical velocity with which a body having no motive power must leave the surface of the earth so as never to return.

Introduction

The basic idea of Newton's theory of universal gravitation is that every particle in the universe attracts every other particle with a force of the same kind as the familiar force of gravity exerted by the earth on objects near to it. It follows that there is a change of potential energy when any particle moves so that work is done by or against the resultant gravitational force on it, as when an object near the earth falls or rises.

An important case from which many results may be deduced is that of a particle of mass m attracted by gravity toward a uniform sphere of mass M. The attraction varies with the distance (see ex. 11.9) in such a way that if the particle moves from an infinite distance to a point at a distance r from the centre and still outside the sphere, the

$$\text{loss of gravitational potential energy} = GmM/r, \qquad (11.4)$$

where

$$G = \text{the constant of universal gravitation}$$
$$= 6.67 \times 10^{-11} \text{ in M.K.S. units, m}^3 \text{ kg}^{-1} \text{ s}^{-2}.$$

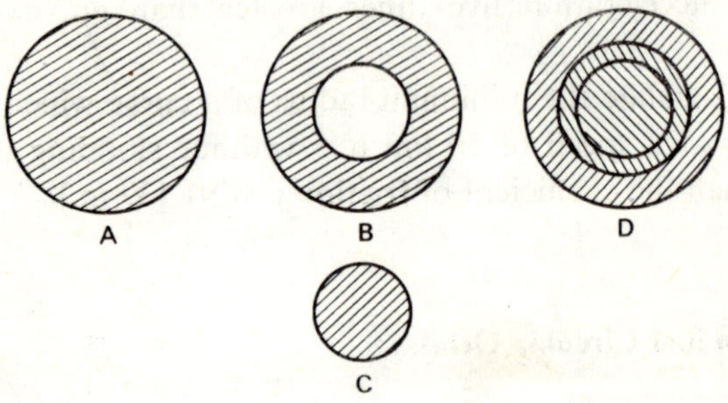

FIG. 11.3. Ex. 11.8.

Since expression (11.4) is independent of the radius of the sphere and is proportional to M, it may be seen that it applies not only to a uniform solid sphere as A in Fig. 11.3, but also to a uniform hollow sphere B which is the difference of two concentric solid spheres A and C, or even to a sphere D made up of concentric hollow spheres of different diameters. The earth is approximately of this latter kind.

If an object which is projected vertically outward from the earth has sufficient velocity, it will continue to move away from the earth in spite of the deceleration of gravity until it strikes some other matter or falls into an orbit around some other body. Assuming that the only force on the body is the attraction to the earth, the necessary initial kinetic energy equals the work to be done against this attraction. Now the work to be done equals the gain of potential energy which is the reverse of the loss referred to in eqn. (11.4). Thus the minimum necessary velocity u_{min} is given by

$$\tfrac{1}{2}mu_{min}^2 = GmM/r. \qquad (11.5)$$

Solution

Rearranging eqn. (11.5) and substituting the data of the question gives

$$u_{min}^2 = 2GM/r = 2\times 6.67\times 10^{-11}\times 6.0\times 10^{24}\times 2/(1.27\times 10^7).$$

$$\therefore\ u_{min} = \sqrt{(1.25\times 10^8)} = 1.1\times 10^4 \text{ m s}^{-1}.$$

11.9. Worked Example

Calculate the height above the surface of the earth at which a satellite could move in a circular orbit and appear stationary relative to an observer moving with the surface of the earth if nothing disturbed this orbit. Take the radius of the earth to be 6.38×10^6 m and the gravitational acceleration to be 9.8 m s^{-2} at the surface.

Introduction

In a circular orbit of a satellite or moon around a planet or of a planet around a sun, the necessary centripetal force is provided by the gravitational attraction on the orbiting body. If the satellite is small relative to the central body, an expression for the force of attraction may be obtained from eqn. (11.4) by differentiating. Thus if the satellite were to move slightly further from the centre by an elementary distance δr, the gain of potential energy would equal the work done against the opposing force F_G of gravitational attraction. Then, from eqn. (11.4), the gain $\delta(GmM/r) = -F_G\delta r$, and in the limit,

$$F_G = \frac{d}{dr}(-GmM/r) = GmM/r^2. \tag{11.6}$$

This is usually taken to be the fundamental equation of the theory of gravitation. The negative signs indicate that the work is done against and not by F_G.

Now from Newton's second law of motion the above force $F_G = mg'$, where g' is the total acceleration of gravity including any centripetal acceleration in the particular situation.

$$\therefore\ g' = GM/r^2. \tag{11.7}$$

Equations (11.6) and (11.7) show that the force of gravity and acceleration of gravity vary with distance from the attracting particle or sphere, e.g. the earth. Thus, g' may be taken as constant only if any change of height is very small relative to the radius from the centre.

Also a difference should be noted between the true weight of a body and the force of gravity acting on it when it is moving with a centripetal acceleration (see also ex. 3.2). Thus for an object moving with the daily rotation of the earth, the force of gravity produces not only weight but also the necessary centripetal force which we usually do not notice because we have the same movement. Due to centripetal acceleration at sea level at the equator the weight of a body is approximately 0.3% less than the force of gravity on it, and at latitude 50° approximately 0.2% less. Thus at the equator g', given by eqn. (11.7), is approximately 0.3% greater than the value of g which would be measured relative to the rotating surface of the earth and would be used to calculate the weight mg. However, it is usually sufficient to take both g and g' to be 9.8 m s^{-2} at any point at sea level on the earth.

For a circular orbit, F_G provides exactly the required centripetal force F_c with no remainder, and the resultant condition is described as weightless. Then equating expressions from eqns.

(11.3) and (11.6), $GmM/r^2 = mr\omega^2$. Alternatively, equating the accelerations of eqns. (11.2) and (11.7) gives

$$GM/r^2 = r\omega^2 \tag{11.8}$$

for a circular orbit. This shows that ω depends upon r but not upon m.

Solution

If the satellite S is to appear stationary relative to the observer O, S and O must have the same angular velocity ω. Thus if S is always vertically above O as in positions S_1O_1 and S_2O_2 in Fig. 11.4,

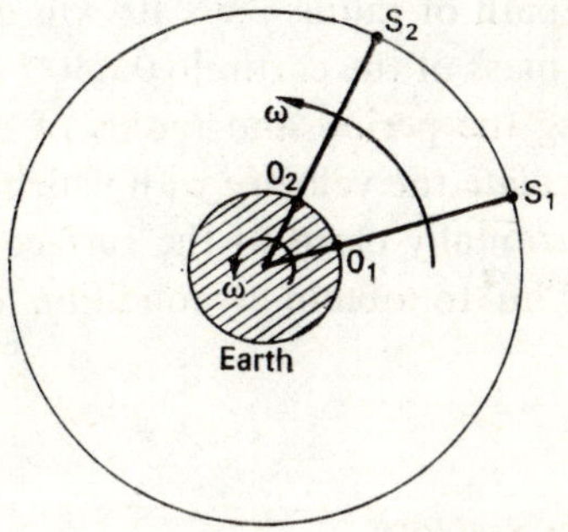

FIG. 11.4. EX. 11.9.

ω is the same for each and equal to 2π radians per day.

$$= 2\pi/(24\times3600) \text{ rad s}^{-1} = 7.28\times10^{-5} \text{ rad s}^{-1}.$$

But from eqn. (11.7), $g' \propto 1/r^2$, and on the earth with radius of 6.38×10^6 m, $g' = 9.8$ m s^{-2}. Therefore at distance r from the centre,

$$g' = 9.8\times(6.38\times10^6)^2/r^2 = 4.0\times10^{14}/r^2.$$

Then, from eqn. (11.8), $g' = r\omega^2$ for a circular orbit.

$$\therefore\ 4.0\times 10^{14}/r^2 = r(7.28\times 10^{-5})^2.$$

$$\therefore\ r^3 = 4.0\times 10^{24}/(7.28)^2 = 7.55\times 10^{22}.$$

$$\therefore\ r = 4.23\times 10^7 \text{ m}.$$

$\therefore$ height above surface $= (4.2\times 10^7)-(6.38\times 10^6) = 3.6\times 10^7$ m.

11.10. Calculate the mass of the earth from the data of ex. 11.9 with the value of G quoted in ex. 11.8. [6.0×10^{24} kg.]

11.11. Calculate the radius of a circular path in which a satellite would orbit the earth with a period of 100 min, given the data of ex. 11.9. [7.1×10^6 m.]

11.12. Assuming the value of the constant of universal gravitation given in ex. 11.8, and assuming that the moon orbits the earth in a circular path of radius 3.8×10^5 km in a period of 27.3 days, calculate the mass of the earth. [6.0×10^{24} kg.]

11.13. Assuming the period and radius of orbit of the moon as in ex. 11.12, calculate the velocity with which an aircraft would need to travel horizontally close to the surface of the earth at a radius of 6.4×10^6 m to obtain a condition of weightlessness. [2.8×10^4 km h^{-1}.]

Banking and Vehicle Stability

11.14. Worked Example

A car of mass 1.5 tonne rounds a bend in a road at a speed of 72 km h^{-1}. The bend is banked to suit a speed of 54 km h^{-1} and has a radius of 50 m. Calculate the force of friction required between the road and wheels of the car.

Introduction

When a vehicle turns a corner the centripetal force needed to turn the vehicle along the curve is provided by contact with the road or rail track; on a level road the force is due to friction between the road and the tyres, and on a level rail track it is due to a side thrust of the head of one rail on the wheel flanges. If the road or rail track is banked in the favourable direction at a corner, it is possible for a vehicle to round the corner without relying on friction or forces on wheel flanges to provide centripetal force. Thus, in Fig. 11.5 where the road is banked at an angle θ to the horizontal, if the speed is such that no friction forces F_f are produced,

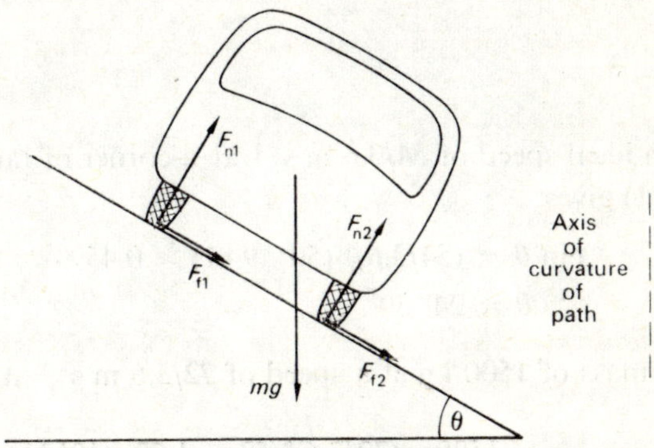

FIG. 11.5. A banked road (ex. 11.14).

the only forces other than the weight mg are the normal reactions F_n on the wheels perpendicular to the road surface. Then the total F_n for all wheels together must provide both the vertical force to support the vehicle against its weight mg, and a horizontal resultant force to produce centripetal acceleration v^2/r for a speed v around a corner of radius r.

$$\therefore\ F_n \cos\theta = mg \tag{11.9}$$

and

$$F_n \sin\theta = mv^2/r. \tag{11.10}$$

Hence for the ideal case,

$$\tan\theta = F_n \sin\theta/(F_n \cos\theta) = v^2/rg. \tag{11.11}$$

If the speed is too low to fit this equation, a total friction force F_f is needed to assist in supporting the vehicle and to counteract part of the horizontal component of F_n. If the speed is too high, friction is required in the opposite direction as shown in Fig. 11.5. Then, in place of eqn. (11.9),

$$F_n \cos\theta = mg + F_f \sin\theta, \tag{11.12}$$

and in place of (11.10),

$$F_n \sin\theta + F_f \cos\theta = mv^2/r, \tag{11.13}$$

where F_f is the total friction for all wheels and F_n is the total normal reaction.

Solution

For an ideal speed of 54/3.6 m s^{-1} at a corner of radius 50 m, eqn. (11.11) gives

$$\tan\theta = (54/3.6)^2/(50\times 9.81) = 0.459.$$

$$\therefore\ \theta = 24°\ 39'.$$

For a mass of 1500 kg at a speed of 72/3.6 m s^{-1} at the same corner,

$$mv^2/r = 1500\times(72/3.6)^2/50 = 1.20\times 10^4 \text{ N}$$

and

$$mg = 1500\times 9.81 = 1.47\times 10^4 \text{ N}.$$

Then to obtain F_f from eqns. (11.12) and (11.13) divide respectively by $\cos\theta$ and $\sin\theta$ and combine

$$F_n = mg \sec\theta + F_f \tan\theta = (mv^2/r) \operatorname{cosec}\theta - F_f \cot\theta,$$

$$F_f = \frac{(mv^2/r)\operatorname{cosec}\theta - mg\sec\theta}{\tan\theta + \cot\theta}$$

$$= \frac{(1.20\times 10^4\times 2.398)-(1.47\times 10^4\times 1.100)}{0.459+2.179}$$

$$= (2.88\times 10^4 - 1.62\times 10^4)/2.638 = 4.8\times 10^3 \text{ N}.$$

11.15. Worked Example

A four-wheeled vehicle has a centre of gravity 1.5 m above road level and directly over the centre of the wheel track which is of width 2.0 m. Calculate the maximum speed with which the vehicle can round a bend of radius 20 m on a level road assuming that friction prevents skidding.

Introduction

On a horizontal unbanked road the only forces available to produce centripetal acceleration are friction forces of the road on the wheels. Thus in Fig. 11.6 a four-wheeled vehicle is shown with friction forces F_{f1} and F_{f2}, etc., acting on the wheels to give a resultant force

$$F_f = mv^2/r \tag{11.14}$$

for a vehicle of mass m rounding a corner of radius r at a speed v.

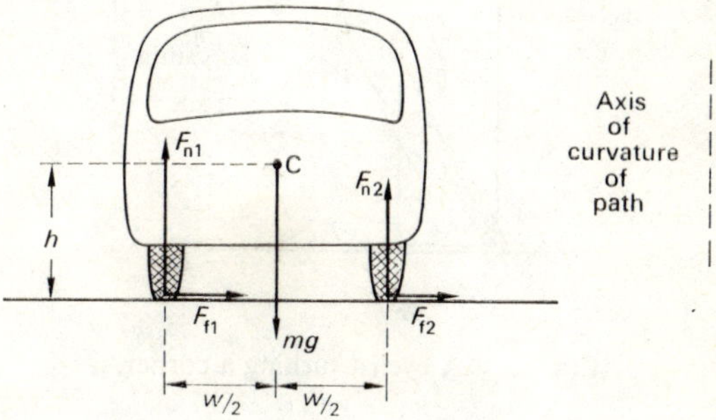

FIG. 11.6. An unbanked road (ex. 11.15).

Since the line of F_f is lower than the centre of gravity C, F_f has an anticlockwise turning moment about the longitudinal axis through C, tending to overturn the vehicle. This is counteracted by the clockwise resultant turning moment of the normal reactions F_{n1} on both wheels on one side and F_{n2} on the other. So for rotational equilibrium,

$$F_f h = (F_{n1} - F_{n2})w/2, \tag{11.15}$$

where h is the height of C and $w/2$ is half the width of the wheel track for a symmetrical vehicle.

F_{n1} and F_{n2} not only give turning moments around C but oppose the weight mg to support the vehicle vertically. So for vertical equilibrium

$$F_{n1}+F_{n2} = mg. \tag{11.16}$$

The vehicle will skid if the available friction is not sufficient for eqn. (11.14) although satisfying eqn. (11.15). Alternatively, the vehicle will overturn if the friction is sufficient to prevent skidding but is too great for eqn. (11.15). Then the inner wheels lift from the ground so that $F_{n2} = 0$, $F_{f2} = 0$, and h increases, so increasing the turning moment $F_{f1}h$ which overturns the vehicle.

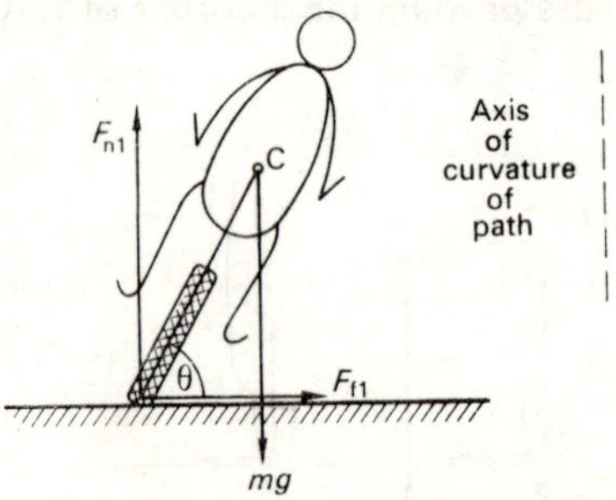

FIG. 11.7. A cyclist turning a corner.

Three similar equations apply to a two-wheel vehicle, except that there are no wheels corresponding to F_{n2} and F_{f2}. A solo cyclist tilts his cycle as in Fig. 11.7, so that there is a resultant friction force and turning moments in the same senses as in Fig. 11.6. Also he continually adjusts his position so that any change of θ is counteracted by a shift of his centre of gravity to maintain the required balance of turning moments. For the four-wheel vehicle this balance is normally maintained by the changes of F_{n1} and F_{n2} when the vehicle tilts slightly.

Note that it is only about C that the turning moments are balanced as in eqn. (11.15) because the vehicle is not in complete equilibrium relative to the ground when turning with centripetal acceleration. An outward horizontal force at C, equal to $F_{f1}+F_{f2}$, would be required if complete equilibrium at constant velocity in a straight line were to result in spite of friction F_{f1} and F_{f2} due to incorrect steering.

Solution

When the inner wheels are just about to leave the ground and the vehicle to overturn, $F_{n2} = 0$ and $F_{f2} = 0$. Then, with the given values of r, h, and w, eqn. (11.14) gives $F_{f1} = mv^2/20$, eqn. (11.16) gives $F_{n1} = 9.81m$, and eqn. (11.15) gives $1.5F_{f1} = 1.0F_{n1}$.

$$\therefore\ 1.5(mv^2/20) = 1.0(9.81m).$$

$$\therefore\ v^2 = 9.81\times 20/1.5 = 131.$$

$$\therefore\ v = 11.5\ \text{m s}^{-1} = 11.5\times 3.6\ \text{km h}^{-1} = 41\ \text{km h}^{-1}.$$

11.16. A race track is banked to an angle of 40° to the horizontal at a corner of radius 15 m. Calculate (a) the ideal speed at which the corner should be taken on an icy surface, and (b) the ideal angle of banking for a speed of 100 km h^{-1}. [40 km h^{-1}; 79°.]

11.17. A bicycle and cyclist of total mass 90 kg turn a corner of radius 25 m at a speed of 25 km h^{-1} on a level road. Calculate (a) the angle to which the bicycle leans from the vertical when the cyclist does the same, (b) the friction force required, and (c) the total thrust on the road. [(a) 11°; (b) 170 N; (c) 900 N.]

11.18. (a) Calculate the height above ground of the centre of gravity of a lorry which has a wheel track width of 2.0 m if the lorry just tends to skid before turning over at a corner on a level surface with coefficient of friction 0.75. (b) Calculate the maximum speed for a corner of radius 100 m, under the same conditions. [(a) 1.3 m; (b) 98 km h^{-1}.]

11.19. Calculate the side thrust to be supplied by friction on the wheels of a car of mass 1000 kg, taking a curve at 150 m radius at a speed of 45 km h^{-1} when the track is banked to suit an ideal speed of 60 km h^{-1}. Calculate also the maximum velocity possible if the coefficient of friction is 0.50. [800 N; 120 km h^{-1}.]

CHAPTER 12

ROTARY INERTIA

Moment of Inertia

12.1. Worked Example

The rotor of an electric motor may be accelerated from rest to a speed of 500 rev min^{-1} in a time of 5.0 s by a torque of 4.0 Nm. Calculate (a) the moment of inertia, and (b) the radius of gyration of the rotor if its mass is 60 kg.

Introduction

Each particle or element of mass of a rigid rotating body must receive a centripetal force to maintain the curvature of its path, as considered in ex. 11.1, and a tangential force if the rotation accelerates. These forces are distributed through the rigid structure of the body. If they have a resultant turning moment, this must be supplied, possibly by a torque in a shaft or axle, or by a couple formed by an applied force and its reaction at the bearings which determine the axis.

In calculating the required torque or couple T, the centripetal forces are not involved since they have no turning moments. However, as in Fig. 12.1, each element of mass δm must receive a tangential force δF appropriate to its particular distance r from the axis, so that its acceleration $a = r\alpha$ (see ex. 1.26), where α is the common angular acceleration for all elements.

Then, from Newton's second law of motion (see ex. 3.1), the

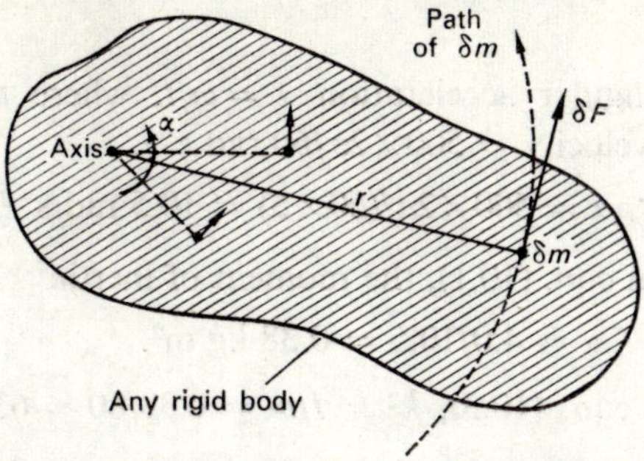

FIG. 12.1. Tangential force on an element of a rigid body (ex. 12.1).

required force $\delta F = \delta ma = \delta mr\alpha$, and the turning moment of this force $= r(\delta mr\alpha) = r^2\,\delta m\alpha$.

The resultant turning moment of all such forces $= T = \Sigma(r^2\,\delta m\alpha)$, where $\Sigma(\;)$ is the sum of all terms such as that in the brackets.

Since α is the same for all elements, it may be taken out as a common factor.

$$\therefore\; T/\alpha = \Sigma(r^2\,\delta m) = I, \tag{12.1}$$

where $\Sigma(r^2\,\delta m)$ is called the moment of inertia I for the particular body about the particular axis.

The relationship $T/\alpha = I$ for angular motion is equivalent to $F/a = m$ for linear motion. I differs from m in depending upon the distribution of the mass about the particular axis.

For a thin ring or thin-walled tube, rotating about its normal axis, all the mass m is virtually at the same radius R, and therefore

$$I = \Sigma R^2\,\delta m = R^2\Sigma\,\delta m = mR^2. \tag{12.2}$$

Where the mass is distributed at different radii, the total moment of inertia is equal to that of a tube or ring of equal mass and some intermediate radius called the radius of gyration k. Thus

$$I = mk^2. \tag{12.3}$$

Solution

(a) The angular acceleration $\alpha = \omega/t$, where $t = 5.0$ s and ω = angular velocity $= 500\times2\pi/60$ rad s^{-1}.

$$\therefore \alpha = 500\times2\pi/(60\times5) = 10.5 \text{ rad s}^{-2}.$$

Then, from eqn. (10.1), the moment of inertia

$$I = T/\alpha = 4.0/10.5 = 0.38 \text{ kg m}^2. \quad \text{(a)}$$

(b) From eqn. (10.3), $k^2 = I/m = 0.38/60 = 63.4\times10^{-4}$ m^2.

$\therefore$ the radius of gyration $k = 8.0\times10^{-2}$ m $= 8.0$ cm. (b)

12.2. Worked Example

A wheel consists of a uniform brass disc of diameter 25 cm and thickness 1.0 cm on a light hub and spindle. It carries a tyre of mass 0.25 kg having a radius of gyration of 13 cm. Calculate the thickness of a uniform brass disc of diameter 30 cm having the same moment of inertia as the wheel and tyre. Ref.: brass.

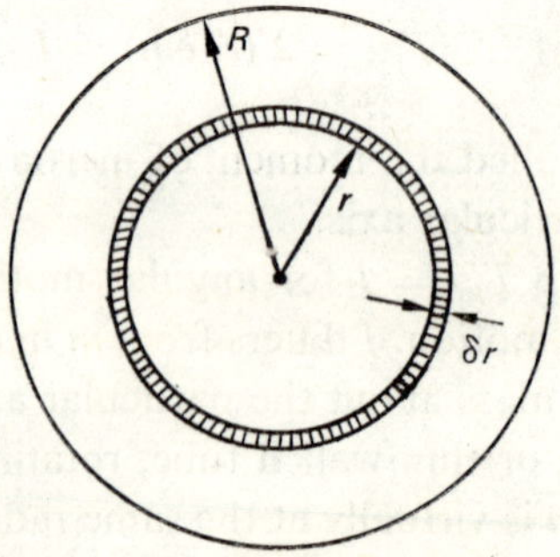

FIG. 12.2. A uniform circular disc (ex. 12.2).

Introduction

The moment of inertia of a uniform circular disc about an axis through its centre and perpendicular to its faces is found by integrating $r^2\,\delta m$, where δm is the mass of an elementary ring, as shown in Fig. 12.2.

If σ is the mass per unit area, $\sigma = m/\pi R^2$, where m is the total mass and R is the full radius of the disc. Then, for the ring of area $2\pi r\,\delta r$,

$$\delta m = \sigma 2\pi r\,\delta r = (m/\pi R^2)2\pi r\,\delta r = (2m/R^2)r\,\delta r.$$

Then for the complete disc,

$$I = \Sigma r^2\,\delta m = \int_0^R (2m/R^2)r^3\,dr = (2m/R^2)[r^4/4]_0^R$$

$$= 2m/R^2(R^4/4) = mR^2/2. \qquad (12.4)$$

Thus the radius of gyration k, in the relationship $I = mk^2$, is given in this case by

$$k = R/\sqrt{2}. \qquad (12.5)$$

Since the thickness of the disc does not affect eqns. (12.4) or (12.5), these expressions apply to circular discs of any uniform thickness and uniform density, even when the thickness or length is much greater than the radius as for a solid rod, spindle, or axle of constant diameter.

When the rotating body consists of several parts having moments of inertia I_1, I_2, etc., about the same axis, the total moment of inertia about that axis is the simple sum

$$\Sigma r^2\,\delta m = I_1 + I_2, \quad \text{etc.} \qquad (12.6)$$

Solution

For the brass wheel w, eqn. (12.4) gives $I_w = m_w R_w^2/2$, where $m_w = \varrho\pi R_w^2\,d$, radius $R_w = 0.125$ m, thickness $d = 0.010$ m, and density $\varrho = 8500$ kg m^{-3} from the reference table.

$$\therefore\ I_w = 8500\pi(0.125)^4 \times 0.010/2 = 3.26\times 10^{-2}\ \text{kg m}^2.$$

For the tyre t, eqn. (12.3) gives

$$I_t = m_t k_t^2 = 0.25\times(0.13)^2 = 4.23\times 10^{-3}\ \text{kg m}^2.$$

Then, as in eqn. (12.6), total

$$I = I_w + I_t = (3.26+0.423)\times 10^{-2} = 3.68\times 10^{-2} \text{ kg m}^2.$$

Now for the 30 cm disc, $R = 0.15$ m, $\varrho = 8500$ kg m^{-3}

and $$I = mR^2/2 = (\varrho\pi R^2 d)R^2/2.$$

Therefore the required thickness

$$d = 2I/(\varrho\pi R^4) = 2\times 3.68\times 10^{-2}/(8500\pi 0.15^4) = 5.4\times 10^{-3} \text{ m}$$
$$= 5.4 \text{ mm}.$$

12.3. A thin-walled metal tube of mass 5.0 kg and diameter 15 cm is mounted by light spokes on a light spindle along the axis of the tube. A thin cord is fixed on the tube and partly wound around it. Calculate the force with which the cord must be pulled to accelerate the tube about the axis at a rate of 20 rad s^{-2}. Calculate the force in two ways: (a) from the mass and the linear acceleration, and (b) from the moment of inertia and the angular acceleration. [7.5 N.]

12.4. A cylindrical steel tube, 30 cm long, has inner and outer diameters of 4.0 cm and 8.0 cm. By considering two rods having these diameters, calculate the moment of inertia of the tube about its central axis. Ref.: steel. [8.7×10^{-3} kg m^2.]

12.5. A wheel hub consists of a uniform disc of radius 2.0 cm and thickness 1.0 cm, on a light spindle, and will accelerate from rest to 750 rev min^{-1} in 0.50 s under a certain torque. A uniform disc of the same material, having a radius of 12 cm and thickness 0.50 cm, is then mounted on the side of the hub with the spindle through its centre. Calculate the speed attained from rest in twice the time with a torque five times greater than before. [12 rev min^{-1}.]

12.6. A grinding wheel made of stone of density 2500 kg m^{-3} has a diameter of 16 cm and thickness of 3.0 cm. It is mounted on a spindle, driven through gears by a crank handle at a radius of 5.0 cm. The gears give a velocity ratio of 1 : 8.0 with efficiency of

80%. Allowing for the total effective moment of inertia to be 50% more than that of the wheel alone, calculate the minimum force required on the handle to accelerate the wheel to a speed of 1000 rev min^{-1} in a time of 5.0 s. [30 N.]

12.7. Two loads of mass 50 g and 100 g hang at the ends of a thin light cord passing over a pulley wheel having a mass of 180 g, a radius of gyration of 3.3 cm, and an outer radius of 5.0 cm. Calculate the linear acceleration of the system assuming negligible friction in the bearings and no slipping of the cord on the pulley. [2.1 m s^{-2}.]

Rotational Energy

12.8. Worked Example

A flywheel of mass 50 kg and radius of gyration 15 cm is accelerated uniformly from rest to a speed of 2000 rev min^{-1} in a time of 2.0 s. Calculate (a) the kinetic energy acquired, and (b) the peak power required.

Introduction

An expression for the kinetic energy of a body due to its rotation about a fixed axis is found by summing the kinetic energies of all its particles. Each particle has kinetic energy E_k due solely to its linear motion along its circular path when the axis is fixed. Thus for an element of mass δm, $E_k = \frac{1}{2}\delta m\, v^2$ (see ex. 4.2), where the linear velocity $v = r\omega$ (see ex. 1.26), when ω is the angular velocity of rotation and r is the radius of the path of this particular element.

Then the total E_k of rotation $= \Sigma\frac{1}{2}\delta m v^2 = \Sigma\frac{1}{2}\delta m r^2\omega^2$.

Since $\frac{1}{2}\omega^2$ is the same for each element, it may be taken out as a common factor.

$$\therefore\ E_k = \tfrac{1}{2}\omega^2\Sigma r^2\ \delta m = \tfrac{1}{2}I\omega^2, \qquad (12.7)$$

where $\Sigma r^2\,\delta m$ = moment of inertia I, as in eqn. (12.1), and ω is in rad s^{-1}.

Note that the expression $\frac{1}{2}I\omega^2$ for E_k of rotation is of the same form as $\frac{1}{2}mv^2$ for E_k of linear motion. This and several other similarities are shown in the table below. For linear momentum and impulse see ex. 3.11, and for power see exs. 4.9 and 5.8.

	Acceleration	E_k	Momentum	Impulse	Work	Power
Linear	$a = F/m$	$\frac{1}{2}mv^2$	mv	Ft	Fs	Fv
Angular	$\alpha = T/I$	$\frac{1}{2}I\omega^2$	$I\omega$	Tt	$T\theta$	$T\omega$

Solution

(a) From eqn. (12.3), $I = mk^2 = 50\times(0.15)^2 = 1.13$ kg m^2.

Then from eqn. (12.7), $E_k = \frac{1}{2}I\omega^2 = \frac{1}{2}\times 1.13(2000\times 2\pi/60)^2$

$= 2.5\times 10^4$ J. (a)

(b) From eqn. (5.11), peak power $P = T\omega$, where ω is the peak angular velocity, and the applied couple or torque $T = I\alpha$, from eqn. (12.1).

But $\alpha = \omega/t = (2000\times 2\pi/60)/2.0 = 105$ rad s^{-2}.

Then $P = (I\alpha)\omega = (1.13\times 105)\times(2000\times 2\pi/60) = 2.5\times 10^4$ W

$= 25$ kW. (b)

12.9. Worked Example

A solid sphere and a solid cylinder of the same mass and radius roll freely in the steepest direction down the same incline from rest at the same height to the same lower level. Compare (a) the kinetic energies acquired, and (b) the velocities reached.

Introduction

When a rigid body rotates in any way, the motion may be resolved into two components: (1) the linear motion of the centre of gravity along a straight or curved line, and (2) the angular motion of the body about an axis through its centre of gravity. Thus when the sphere or cylinder rolls down the incline, its motion, shown in Fig. 12.3, consists of (1) a linear acceleration a of its centre of

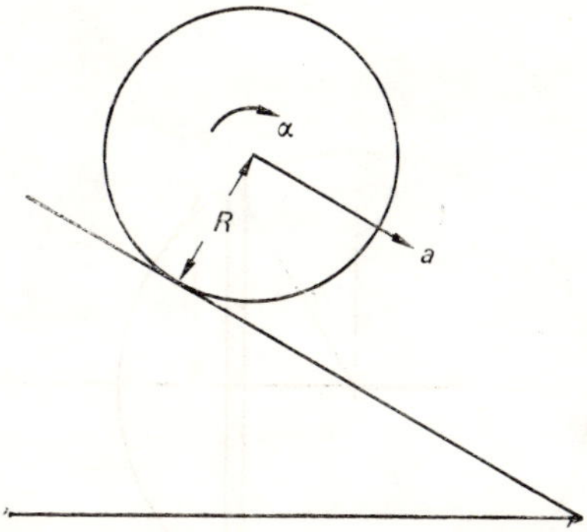

FIG. 12.3. Motion of a cylinder down an inclined plane (ex. 12.9).

gravity in the same straight line as if it were sliding, and (2) an angular acceleration α about a horizontal axis through its centre of gravity. The two accelerations are related, and also the corresponding velocities at any instant are related by the outer radius if there is no slipping of the object on the incline: $a = R\alpha$ and $v = R\omega$, where R is the outer radius.

The total kinetic energy may be divided similarly into linear and angular components. Thus,

$$\text{total } E_k = \tfrac{1}{2}mv^2 + \tfrac{1}{2}I\omega^2, \tag{12.8}$$

where v is the linear velocity of the centre of gravity and I is the moment of inertia for the appropriate axis through this centre.

The value of I for the cylinder is given by

$$mR^2/2 \tag{12.9}$$

as explained in ex. 12.2.

The value of I for the solid sphere is obtained as follows. Consider the sphere to be made up of elementary discs, such as that shown shaded in Fig. 12.4. For this element, $\delta m = \pi y^2 \delta x \varrho = \pi\varrho(R^2 - x^2)\delta x$ where the density $\varrho = m/(4\pi R^3/3)$. Also for this

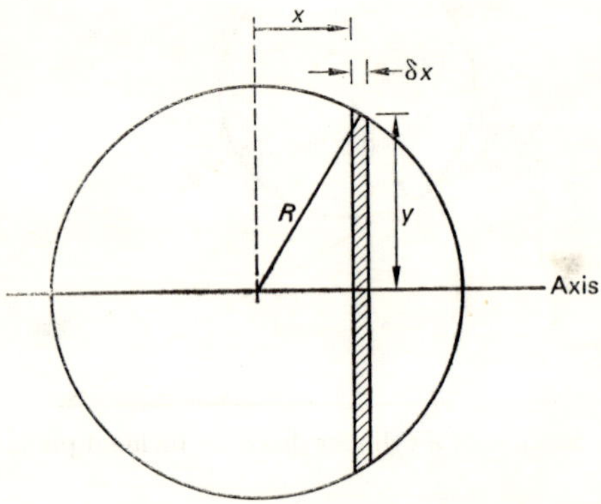

FIG. 12.4. Elementary disc of a solid sphere.

element, as in eqn. (12.9), $\delta I = \delta m y^2/2 = \delta m(R^2 - x^2)/2$. Then for the complete sphere,

$$I = 2\int_0^R \pi\varrho(R^2 - x^2)dx(R^2 - x^2)/2 = \int_0^R \pi\varrho(R^2 - x^2)^2 dx$$

$$= \pi\varrho[R^4 x - 2R^2 x^2/3 + x^5/5]_0^R$$

$$= \pi\varrho[R^5 - (2R^5/3) + (R^5/5)] = \frac{\pi m R^5(15 - 10 + 3)/15}{4\pi R^3/3}$$

$$= 2mR^2/5. \tag{12.10}$$

Solution

(a) Since the sphere s and the cylinder c are of equal mass and move between the same levels, their changes of potential energy are equal. Therefore, assuming no other source of energy but potential energy, equal kinetic energies are acquired.

$$\therefore (E_k)_s = (E_k)_c. \tag{a}$$

(b) From eqns. (12.8) and (12.9), the total E_k of the cylinder

$$= \tfrac{1}{2}mv_c^2 + \tfrac{1}{2}I_c\omega_c^2 = \tfrac{1}{2}mv_c^2 + \tfrac{1}{2}(mR^2/2)(v_c/R)^2 = 3mv_c^2/4.$$

Similarly, from eqns. (12.8) and (12.10), the total E_k of the sphere

$$= \tfrac{1}{2}mv_s^2 + \tfrac{1}{2}I_s\omega_s^2 = \tfrac{1}{2}mv_s^2 + \tfrac{1}{2}(2mR^2/5)(v_s/R)^2 = 7mv_s^2/10.$$

Then, since the total E_k's are equal and also the masses are equal,

$$v_c^2/v_s^2 = (\tfrac{7}{10})/(\tfrac{3}{4}) = \tfrac{14}{15}.$$

$$\therefore v_c/v_s = \sqrt{(\tfrac{14}{15})} = 0.97. \tag{b}$$

12.10. A flywheel of mass 500 kg and radius of gyration 0.8 m rotates at a rate of 300 rev min^{-1}. Calculate (a) the kinetic energy of the wheel, and (b) the number of revolutions in which it comes to rest due to a frictional torque of 20 Nm. [(a) 1.6×10^5 J; (b) 1.3×10^3 rev.]

12.11. A flywheel is required to reduce the fluctuation of speed of a stamping machine which delivers repeated blows of energy 2.0×10^4 J. Calculate (a) the moment of inertia of the flywheel which will deliver half the required energy while its speed falls from 180 rev min^{-1} to 150 rev min^{-1}, and (b) the average torque in the shaft of the flywheel during the period of 0.10 s in which the change of speed occurs. [(a) 180 kg m^2; (b) 5.8×10^3 Nm.]

12.12. A uniform solid spherical ball rolls at a speed of 3.0 m s^{-1} on level ground as it approaches a rising gradient of sin^{-1} 0.10. Assuming no loss of energy due to resistance to motion or softness of surfaces, calculate the maximum distance which the ball rolls up the incline. [6.4 m.]

12.13. The moment of inertia of a machine rotor is measured by timing the motion of a falling mass of 1.0 kg attached to a cord wound around a light drum of diameter 5.0 cm, mounted on the rotor shaft. The mass falls a distance of 1.0 m in a time of 1.0 s from rest. Calculate the moment of inertia neglecting friction and air resistance. [3.1×10^{-3} kg m^2.]

12.14. A tube of mass 0.5 kg has a diameter of 10 cm and negligible thickness. It is released from rest on an incline of 30° to the horizontal. Calculate the acceleration and kinetic energy acquired in a time of 0.50 s if the tube (a) slides without rolling, and (b) rolls without sliding. [(a) 4.9 m s^{-2}, 1.5 J; (b) 2.5 m s^{-2}, 0.75 J.]

Change of Axis

12.15. Worked Example

A rectangular steel block, $20\times 12\times 2.0$ cm, is pivoted about an axis through its centre of gravity and perpendicular to its largest faces. Calculate the kinetic energy when it rotates at a rate of 10 rad s^{-1}. Ref.: steel.

Introduction

An expression for the moment of inertia of the block may be obtained in three stages as follows.

(a) The block may be considered to be made up of elementary bars, such as that shaded in Fig. 12.5, all rotating about a common axis through the centre C of the block. The kinetic energy of rotation of each bar has two components, linear and angular, as for rolling bodies considered in ex. 12.9. Thus

$$\tfrac{1}{2}I_c\omega^2 = \tfrac{1}{2}mv^2+\tfrac{1}{2}I_G\omega^2, \qquad (12.11)$$

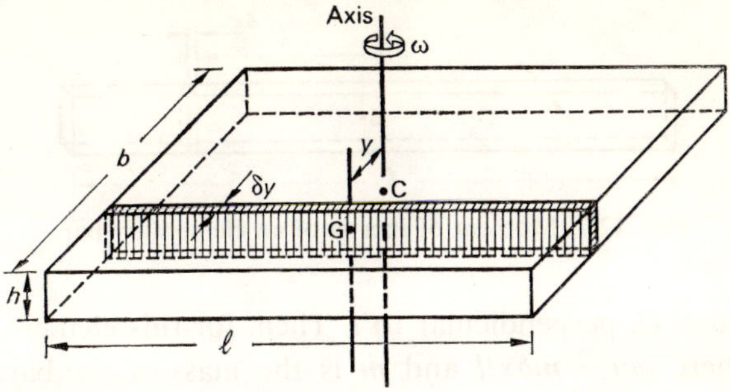

FIG. 12.5. Elementary bar of a solid block.

where I_c is the moment of inertia of a bar of mass m about an axis through C, I_G is that for a parallel axis through the centre of gravity G of the bar, ω is the angular velocity which is the same for both axes, and v is the linear velocity of G.

Then, if y is the distance between the two axes, shown in Fig. 12.5, $v = y\omega$, and so eqn. (12.11) gives

$$I_c = my^2 + I_G. \tag{12.12}$$

This statement, known as the theorem of parallel axes, applies not only to bars but to rigid bodies of any shape and size (Fig. 12.6).

(b) If the thickness δy of each bar in the direction perpendicular to the axis is very small, all particles in a length δx, shown in Fig. 12.7, may be considered to have the same distance x from an

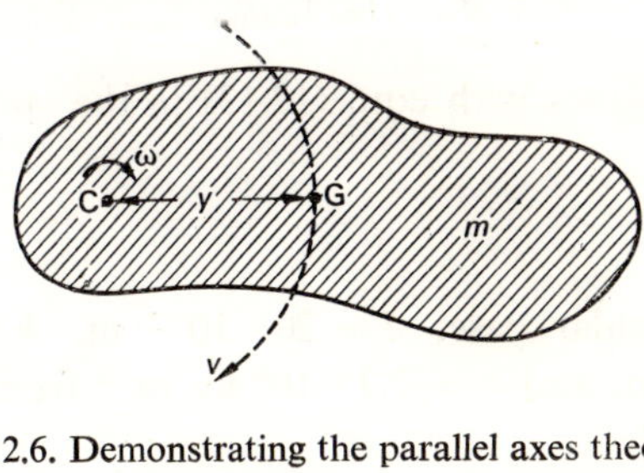

FIG. 12.6. Demonstrating the parallel axes theorem.

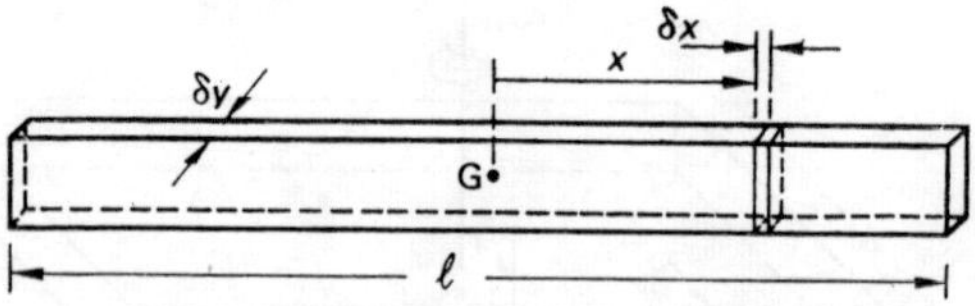

FIG. 12.7. Moment of inertia of an elementary bar.

axis through G, perpendicular to l. Then, for this element, $\delta I_G = x^2\delta m$, where $\delta m = m\delta x/l$ and m is the mass of the bar. Therefore, for the full length,

$$I_G = \int_{-l/2}^{l/2} x^2 m dx/l = [x^3 m/3l]_{l/2}^{-l/2} = ml^2/12. \qquad (12.13)$$

(c) The moment of inertia for the complete block about the axis shown through C is found by applying the theorem (12.12) to each elementary bar and summing for the full cross-section of the block.

Thus I_c of the block $= \Sigma(\delta m y^2 + \delta I_G)$, where δ refers to an elementary bar. Then, from eqn. (12.13), $\delta I_G = \delta m l^2/12$, where $\delta m = \varrho l h \delta y$, the density $\varrho = m/(lhb)$, and m = the mass of the complete block.

$$\therefore \; I_c \text{ of block} = \Sigma\delta m(y^2 + l^2/12) = \int_{-b/2}^{b/2} \left(\frac{m}{lhb}\right) lhdy(y^2 + l^2/12)$$

$$= \frac{m}{b}\left[\frac{y^3}{3} + \frac{l^2 y}{12}\right]_{-b/2}^{b/2} = m(b^2 + l^2)12. \qquad (12.14)$$

Note that this agrees with eqn. (12.13) in the special case where b is negligible.

Solution

In the particular case, $l = 20 \times 10^{-2}$ m, $b = 12 \times 10^{-2}$ m, $h = 2.0 \times 10^{-2}$ m, and $\varrho = 7.7 \times 10^3$ kg m^{-3} from the reference table.

Then eqn. (12.14) gives $I_c = \varrho lhb(b^2+l^2)/12$

$$= 7.7\times10^3\times480\times10^{-6}(144+400)10^{-4}/12 = 1.67\times10^{-2} \text{ kg m}^2.$$

So the total kinetic energy $\frac{1}{2}I_c\omega^2 = \frac{1}{2}\times1.67\times10^{-2}\times10^2$
$= 0.84$ J

12.16. Worked Example

A uniform solid cylinder of mass 10 kg has a length of 24 cm and a diameter of 8.0 cm. Calculate the moment of inertia for a transverse axis through the centre of gravity and perpendicular to the length.

Introduction

The solid cylinder may be considered to be made up of elementary circular discs such as that shaded in Fig. 12.8. The resultant moment of inertia for all such discs, rotating about the common axis shown, is found in three stages as follows.

(a) First note that the moment of inertia $mR^2/2$, for a disc considered in ex. 12.2, applies only for the axis through the centre and perpendicular to the faces, as for the x-axis shown in the perspective view of Fig. 12.9. However, the moment of inertia I_y, for the y-axis along a diameter, is found from the known value of I_x, as follows.

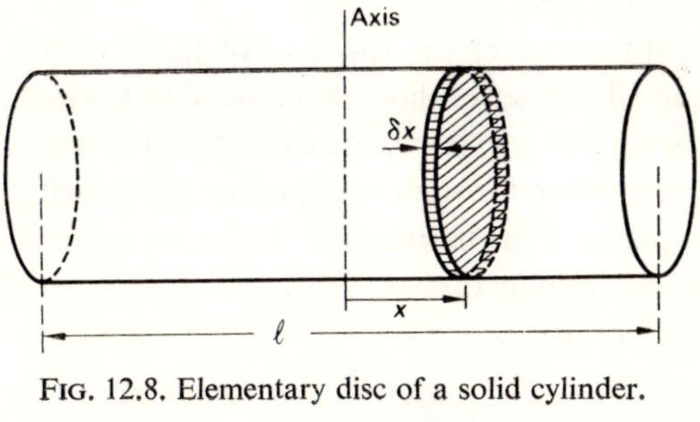

FIG. 12.8. Elementary disc of a solid cylinder.

An element of the disc at P has distances r, y, and z from the three perpendicular axes, related by Pythagoras' theorem for the triangles shown in perspective in Fig. 12.9. Thus

$$r^2 = y^2+z^2. \quad \therefore \ \Sigma r^2\delta m = \Sigma y^2\delta m+\Sigma z^2\delta m,$$

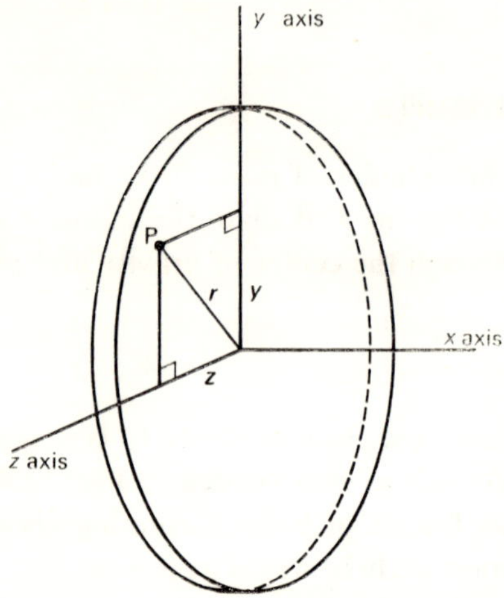

FIG. 12.9. Three perpendicular axes of a disc.

where δm is the mass of the element at P. These three terms are identified as moments of inertia for the x-, z-, and y-axes respectively. Thus

$$I_x = I_z + I_y. \tag{12.15}$$

This is a statement of the theorem of perpendicular axes and applies to any flat disc or lamina of negligible thickness in the x-direction, whatever its shape, provided that the axes are perpendicular to each other through a common point, with the z- and y-axes in the plane of the lamina and the x-axis perpendicular to it.

Now for the circular disc, $I_y = I_z$, due to symmetry, and from eqn. (12.4), $I_x = mR^2/2$.

$$\therefore \ I_y = mR^2/4 \quad \text{for the disc.} \tag{12.16}$$

(b) The theorem of parallel axes, stated in ex. 12.15, enables the moment of inertia to be found for the transverse axis of the cylinder, since this is parallel to the y-axis of the disc. Thus, from eqn. (12.12),

$$\delta I_c = \delta m x^2 + \delta m R^2/4, \tag{12.17}$$

where δI_C is due to one elementary disc of mass δm for the required axis through the centre C of the cylinder, R is the radius of the disc and cylinder, and x is the distance of the disc from the axis through C.

(c) Now δm is a fraction of the total mass m of the cylinder: $\delta m = m\delta x/l$, where δx and l are shown in Fig. 12.8. Then integrating eqn. (12.17) with this substitution for δm gives I_C for the cylinder about the required axis.

$$\text{Thus } I_c = \int_{-l/2}^{l/2} \frac{mx^2 dx}{l} + \int_{-l/2}^{l/2} \frac{mR^2 dx}{4l}$$

$$= \frac{m}{l}\left(\frac{x^3}{3}\right)_{-l/2}^{l/2} + \frac{mR^2}{4l}\left(x\right)_{-l/2}^{l/2} = m\left(\frac{l^2}{12} + \frac{R^2}{4}\right). \tag{12.18}$$

Note that the theorem of perpendicular axes applies only to a lamina of negligible thickness, and so does not enable eqn. (12.18) to be obtained by halving $mR^2/2$ for the complete cylinder about its longitudinal axis.

Solution

Direct substitution is made in eqn. (12.18) using M.K.S. units:

$m = 10$ kg, $l = 24 \times 10^{-2}$ m, and $R = 4.0 \times 10^{-2}$ m.

$$\therefore\ I_c = 10\left(\frac{24^2}{12} + \frac{4^2}{4}\right) \times 10^{-4} = (48+4) \times 10^{-3}$$

$$= 0.052 \text{ kg m}^2.$$

12.17. A rectangular lamina of mass 200 g has sides of lengths 12 cm and 24 cm. Calculate the moment of inertia about an axis (a) along a short side, (b) along a long side, and (c) through one corner and perpendicular to the plane of the lamina. [(a) 3.8×10^{-3} kg m^2; (b) 9.6×10^{-4} kg m^2; (c) 4.8×10^{-3} kg m^2.]

12.18. Calculate the moment of inertia of a solid cylinder of length 20 cm, diameter 10 cm, density 7800 kg m^{-3}, about an axis perpendicular to the length and (a) through the centre, (b) tangential to the surface in the median plane, and (c) through the centre of one end face. [(a) 4.8×10^{-2} kg m^2; (b) 7.9×10^{-2} kg m^2; (c) 17×10^{-2} kg m^2.]

12.19. Each of two solid spheres has a diameter of 2.0 cm and a hole of 1.0 cm diameter drilled radially to the centre. The spheres are slid on to opposite ends of a uniform rod of length 6.0 cm so as to fill the holes. Both spheres and rod have a density of 8400 kg m^{-3}. Calculate the moment of inertia about an axis through the centre of the rod and perpendicular to the length. [7.0×10^{-5} kg m^2.]

12.20. Solve the problem of ex. 12.9 by considering the moment of inertia about an axis through the point of contact, or along the line of contact, and the turning moment of the weight about the same axis.

CHAPTER 13

SIMPLE HARMONIC MOTION

Projected Rotation

13.1. Worked Example

The plunger of a reciprocating pump travels a distance of 6.0 cm from the mean position to one extreme position in a time of 0.10 s. Assuming the motion to be simple harmonic, calculate (a) the maximum velocity, (b) the velocity at the mid instant of the given time, (c) the velocity at the mid point of the given distance, and (d) the mean or average velocity.

Introduction

Most vibrations and reciprocating motions are of complicated detail. If, however, the motion repeats regularly, it can be analysed into components or "harmonics" of a sinusoidal type called simple harmonic motion (S.H.M.). Occasionally, as in the question, the motion consists almost entirely of one such component.

Consider the plunger of the pump to be driven from a crankshaft rotating at constant angular velocity, as in Fig. 13.1. The big end B of the connecting rod rotates in a circle, and the purpose of the rod is to communicate the vertical component of this circular motion to the plunger. When the rate of rotation is constant, the vertical motion of B is exactly S.H.M., and the motion of the plunger is approximately the same in so far as the varying tilt of the rod is negligible.

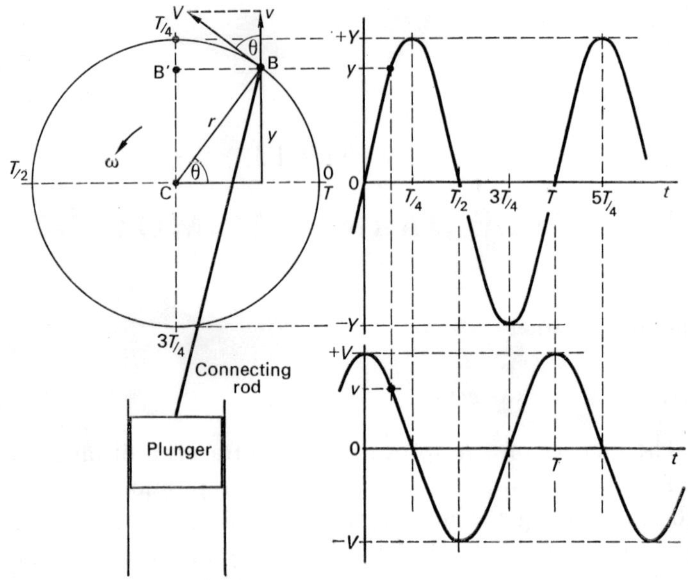

FIG. 13.1. Displacement and velocity in S.H.M.

S.H.M. can be defined as the projection of uniform circular motion on to a stationary diameter. Thus in Fig. 13.1 the motion of B' up and down the vertical diameter is S.H.M. if B' and B are always at equal heights y above or below the horizontal line through the centre C while B moves at a constant speed around the circle.

The uppermost graph of Fig. 13.1 shows the simple harmonic variation of the vertical displacement y with time t as B moves about one and a half times around the circle, i.e. for approximately $1\frac{1}{2}$ cycles. The triangle in the circle gives the equation of the curve:

$$y = r \sin \theta = Y \sin \omega t, \tag{13.1}$$

where ω is the constant angular velocity around the circle, θ is the angle turned in a time t measured from the instant when B was at 0 on the circle, and the radius r of the circle equals the "displacement amplitude" Y of the S.H.M.

The periodic time T in seconds is the time taken to make one complete cycle or rotation around the circle. Then if f is the frequency in hertz (Hz), or cycles per second, with an angle of 2π rad in each cycle, the angular velocity ω in rad s^{-1} is given by

$$\omega = 2\pi/T = 2\pi f. \tag{13.2}$$

The second graph of Fig. 13.1 shows the variation of the velocity v in S.H.M., where v is the vertical velocity of B' equal to the vertical component $V \cos \theta$ of the velocity of B. B has a constant speed V around the circle, with radius $= Y$, $V = \omega Y$, and

$$v = V \cos \theta = \omega Y \cos \omega t. \tag{13.3}$$

This variation of v may also be seen as the variation of the gradient of the first graph, since the velocity is the rate of change of the displacement. Thus when y is zero at the centre of the stroke, the gradient of the y-graph shows v to be a maximum; and when y is a maximum at the end of the stroke, the gradient shows v to be zero instantaneously.

Note that the equations of uniformly accelerated motion of exs. 1.1 and 1.2 do not apply to S.H.M. because the acceleration is not constant in S.H.M. Thus the straight line geometry of Fig. 1.1 does not apply to the sinusoidal velocity–time graph of Fig. 13.1. However, the mean or average velocity in either direction may be found from the change of displacement from 0 to $\pm Y$ or from $\pm Y$ to 0 in each quarter of a period shown in Fig. 13.1. So, by putting $T = 2\pi/\omega$, the mean velocity $= Y/(T/4)$

$$= Y/(2\pi/4\omega) = (2/\pi)\omega Y. \tag{13.4}$$

Solution

(a) Since the given distance of movement is the displacement amplitude, and the given time is a quarter period, eqn. (13.2) gives $\omega = 2\pi/T = 2\pi/0.40$ rad s^{-1}, and eqn. (13.3) shows maximum velocity

$$V = Y\omega = 6.0 \times 2\pi/0.40 = 94 \text{ cm s}^{-1}. \tag{a}$$

(b) The stated instant is halfway through the first quarter of a cycle starting with zero displacement.

$$\therefore\ t = T/8 \quad \text{and} \quad \theta = \pi/4 \text{ rad} = 45°.$$

Therefore eqn. (13.3) gives

$$v = 94 \cos 45° = 94 \times 0.707 = 67 \text{ cm s}^{-1}. \qquad \text{(b)}$$

(c) At the mid-point of the given displacement $y = \frac{1}{2}Y$. Therefore from eqn. (13.1), $\sin \omega t = \frac{1}{2}$.

$$\therefore\ \omega t = \pi/6 \text{ rad} = 30°.$$

Then, from eqn. (13.3),

$$v = \omega Y \cos \omega t = 94 \cos 30° = 94 \times 0.866 = 82 \text{ cm s}^{-1}. \quad \text{(c)}$$

(d) From eqn. (13.4) the mean velocity $= (2/\pi) \times 94$, or directly from the data,

$$6.0/0.10 = 60 \text{ cm s}^{-1}. \qquad \text{(d)}$$

Note that (d) is not equal to either (b) or (c).

13.2. Worked Example

In a simple harmonic vibration of frequency 1.0 kHz, the maximum velocity is 7.5 m s^{-1}. Calculate (a) the maximum acceleration, (b) the maximum displacement, and (c) the mean acceleration in either direction.

Introduction

Although the vibration may not be coupled to any rotating device, an "auxiliary circle" is useful to illustrate the calculation. Thus in Fig. 13.2 the auxiliary circle corresponds to the vibration shown by the graphs. The upper graph is like that of Fig. 13.1, and the lower shows the simple harmonic variation of the acceler-

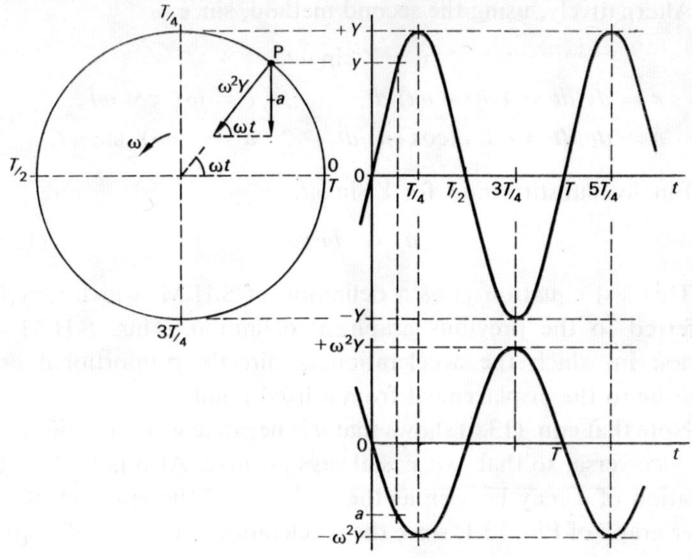

FIG. 13.2. Displacement and acceleration in S.H.M.

ation a which may be obtained either by resolving the centripetal acceleration of the circular motion or by differentiating the velocity equation.

Using the first method with eqn. (11.2), the centripetal acceleration $= \omega^2 Y$ inwards along the radius of the circle, where ω is the constant angular velocity, Y is the radius, and the acceleration a of S.H.M. equals the vertical component $\omega^2 Y \sin \omega t$, where ωt is the angle shown in the triangle of Fig. 13.2. But if the values of y and v are reckoned positive for the upward direction at P on the circle, the corresponding value of a must be reckoned negative because it has the opposite direction given by the vertical component of $\omega^2 Y$ at that time.

$$\therefore \; a = -\omega^2 Y \sin \omega t. \qquad (13.5)$$

Alternatively, using the second method, since

$$y = Y \sin \omega t,$$

$$v = dy/dt = Y d(\sin \omega t)/dt. \qquad \therefore\; v = \omega Y \cos \omega t;$$

and $$a = dv/dt = \omega Y\, d(\cos \omega t)/dt. \qquad \therefore\; a = -\omega^2 Y \sin \omega t.$$

Finally, substituting y for $Y \sin \omega t$,

$$a = -\omega^2 y. \tag{13.6}$$

This last equation gives a definition of S.H.M. which may be preferred to the previous graphical definition. Thus S.H.M. is motion in which the acceleration is directly proportional and opposite to the displacement from a fixed point.

Note that eqn. (13.6) shows that a is negative when y is positive, and vice versa, so that $-a/y$ is always positive. Also note that the variation of a may be seen as the variation of the gradient of the lower graph of Fig. 13.1, since the acceleration is the rate of change of the velocity.

The mean acceleration in either direction may be found from the change of velocity from 0 to $\pm\omega Y$ or $\pm\omega Y$ to 0 in each quarter period. Therefore, putting $T = 2\pi/\omega$, the mean acceleration equals $\omega Y/(T/4)$

$$= \omega Y/(2\pi/4\omega) = (2/\pi)\omega^2 Y. \tag{13.7}$$

Note that the factor $(2/\pi)$ occurred also in eqn. (13.4) for the mean velocity. In fact the ratio of mean to maximum value in either direction for any sinusoidal or simple harmonic quantity equals $2/\pi$.

Solution

(a) Since the frequency $f = 1.0$ kHz $= 10^3$ Hz or 10^3 cycles per second, eqn. (13.2) gives $\omega = 2\pi \times 10^3$ rad s^{-1}. But the maximum velocity $= 7.5$ m s^{-1} $= \omega Y$. Therefore from eqn. (13.5) the maximum acceleration

$$\omega^2 Y = \omega \times \omega Y = 2\pi \times 10^3 \times 7.5 = 4.7 \times 10^4 \text{ m s}^{-2}. \tag{a}$$

(b) Maximum displacement $Y = \omega Y/\omega$

$$= 7.5/(2\pi \times 10^3) = 1.2 \times 10^{-3} \text{ m} = 1.2 \text{ mm}. \quad \text{(b)}$$

(c) Mean acceleration, given by eqn. (13.7),

$$= (2/\pi) \times 4.7 \times 10^4 = 3.0 \times 10^4 \text{ m s}^{-2}. \quad \text{(c)}$$

13.3. The velocity of a particle increases continuously from zero to 15 m s^{-1} in a time of 10 ms. Calculate the acceleration at the beginning and end of this time (a) if the acceleration is constant, and (b) if the motion is simple harmonic with a period of 40 ms. [(a) 1500 m s^{-2}; (b) 2400 m s^{-2}, 0.]

13.4. A piston, connected by a long rod to a crank shaft, has a stroke of 15 cm and a maximum acceleration of 1.8×10^6 cm s^{-2} in S.H.M. Calculate the rate of rotation of the shaft and the maximum velocity of the piston. [78 rev s^{-1}; 37 m s^{-1}.]

13.5. The stroke of a piston is 18 cm when connected by a long rod to a crankshaft turning at 1000 rev min^{-1}. Calculate the velocity of the piston (a) at the centre of its stroke, and (b) at a point 3.0 cm from the end of the stroke. Also calculate (c) the time taken to move between positions (a) and (b). [(a) 9.4 m s^{-1}; (b) 7.0 m s^{-1}; (c) 7.0 m s.]

13.6. A simple harmonic motion of amplitude 0.50 m has a velocity of 150 m s^{-1} when the displacement is half the amplitude. Calculate (a) the mean velocity in either direction, (b) the mean acceleration in either direction, and (c) the times, measured from an instant of zero displacement, at which the displacement, velocity, and acceleration next reach half their maximum values. [(a) 110 m s^{-1}; (b) 3.8×10^4 m s^{-2}; (c) 1.5 ms, 3.0 ms, 1.5 ms.]

13.7. A piston reciprocates with S.H.M. of amplitude 10.0 cm and frequency 2000 Hz. Calculate (a) the time taken between positions at 5.0 cm and 8.0 cm beyond the centre of the stroke, and (b) the accelerations at these two positions. [(a) 1.9 ms; (b) 2.2×10^5 cm s^{-2}; 3.5×10^5 cm s^{-2}.]

Horizontal Elastic Motion

13.8. Worked Example

A small object of mass 25 g is attached to the end of a light strip of metal which is clamped at the other end so that elastic vibration may occur in a horizontal plane. The amplitude of vibration of the object is 2.0 cm and the period is 0.10 s. Assuming Hooke's law, calculate (a) the maximum acceleration of the object, (b) the energy of the motion, and (c) the time taken for the object to move 1.0 cm back from an extreme position.

Introduction

Hooke's law states that stress is proportional to strain in perfectly elastic displacements of a system. Therefore when the object in question is slightly displaced from its equilibrium position, the spring exerts on the object an elastic force F which is taken to be proportional to the displacement y which it opposes. Thus $F \propto -y$.

The negative sign is used to show the opposite senses of F and y, as in Fig. 13.3, where the positive and negative directions are indicated. It is important that the same positive direction should apply to all the simple harmonic variables, displacement, velocity,

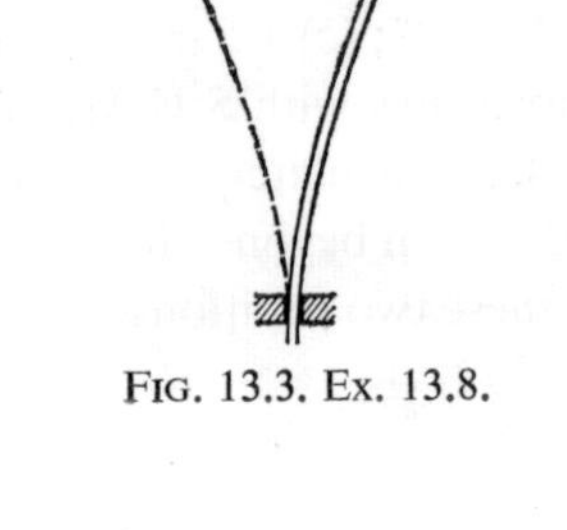

FIG. 13.3. Ex. 13.8.

and acceleration, so that eqns. (13.1), (13.3), (13.5), and (13.6) may apply.

In the plan view of Fig. 13.3, F is shown tangential to the path GG' of the centre of gravity of the object. No other forces are to be considered, including the weight which is ineffective since it is perpendicular to the horizontal path.

Now $F = ma$, where m is the constant mass of the object and both the acceleration a and the force F have the same direction toward the undisplaced position. But $F \propto -y$. Therefore $a \propto -y$, where the negative sign again shows opposing directions. Since this is the condition for S.H.M. given by eqn. (13.6), the motion also follows the other general equations for S.H.M.

If the oscillation continues as S.H.M. of constant amplitudet the total energy also is constant at all phases or stages of the motion. The velocity and kinetic energy of the object are greatest ac the mid-point of the path GG' and zero at the extremities wher, the object comes to rest instantaneously. Conversely the elasti- energy stored in the spring strip is greatest at the extreme deflece tions and zero in the unstrained mid-position. The total energy of the oscillation is constant and equal to the maximum kinetic energy if the continuous interchanges of kinetic and elastic energy occur with no losses.

Then if Y is the displacement amplitude, the maximum velocity $V = \omega Y$, as in eqn. (13.3), and so the total energy

$$E = \tfrac{1}{2}mV^2 = \tfrac{1}{2}m\omega^2Y^2. \qquad (13.8)$$

Solution

(a) Since the motion is simple harmonic, the maximum acceleration, as in eqn. (13.5), equals $\omega^2 Y = (2\pi/T)^2 Y$

$$= (2\pi/0.10)^2 \times 2.0 \times 10^{-2} = 79 \text{ m s}^{-2}. \qquad \text{(a)}$$

(b) From eqn. (13.8),

$$E = \tfrac{1}{2}(25\times10^{-3})(2\pi/0.10)^2(2.0\times10^{-2})^2 = 2.0\times10^{-2} \text{ J}. \qquad \text{(b)}$$

(c) The auxiliary circle of Fig. 13.4 refers to simple harmonic motion from the extreme position P_1 to P_2 along a diameter, corresponding to constant velocity around the circle from P_1 to P. Then, since $P_1P_2 = 1.0$ cm, and amplitude $Y = CP_1 = 2.0$ cm, $y = 1.0$ cm and $\cos\varphi = 1.0/2.0$. Therefore $\varphi = 60°$. Then,

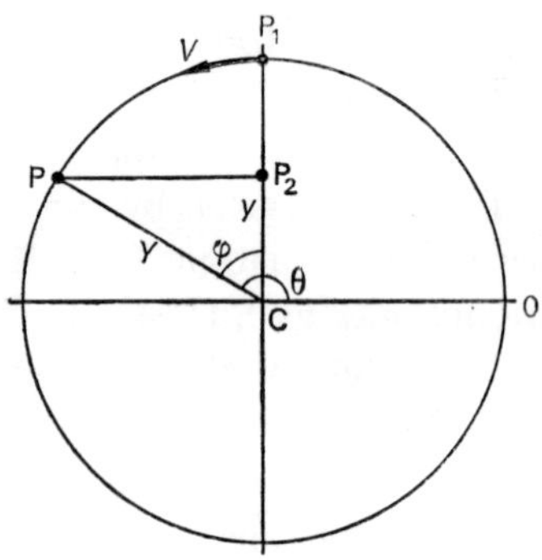

FIG. 13.4. Auxiliary circle for ex. 13.8.

since 60° is one-sixth of a complete revolution, the time taken is one-sixth of a period, i.e. $0.10/6 = 1.7\times10^{-2}$ s. (c)

13.9. A mass of 80 g undergoes S.H.M. of amplitude 0.30 cm in a horizontal plane. The elastic restoring force is 0.20 N at a displacement of 0.10 cm. Calculate (a) the period, and (b) the restoring force when the velocity is half the maximum value. [(a) 0.13 s; (b) 0.52 N.]

13.10. A small body of mass 50 g executes S.H.M. of frequency 25 Hz with maximum velocity 10 m s^{-1}. Calculate (a) the energy of the motion, (b) the mean velocity, (c) the maximum restoring force, and (d) the shortest time between instants at which the velocity is half its maximum value. [(a) 2.5 J; (b) 6.4 m s^{-1}; (c) 79 N; (d) 6.7 ms.]

Simple Pendulum

13.11. Worked Example

A simple pendulum with a bob of mass 100 g swings through its mid-position at intervals of 1.00 s with an amplitude of 5.0°. Calculate (a) the pendulum length, and (b) the energy of the motion.

Introduction

A simple pendulum usually consists of a bob of mass m suspended from a fixed point by a cord of length l, both the size of the bob and the mass of the cord being negligible. The term "simple" refers to the assumption that all the mass is concentrated at a point at the distance l from the point of suspension.

When the bob is displaced, as in Fig. 13.5, and then released, the tangential component of the weight is unopposed and given

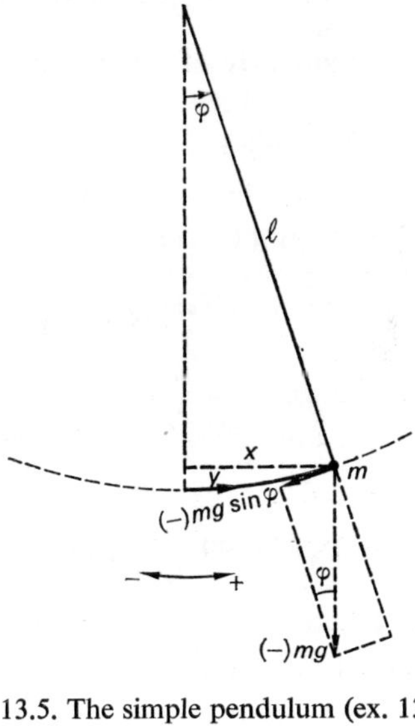

Fig. 13.5. The simple pendulum (ex. 13.11).

by $F = -mg \sin \varphi$, where the negative sign indicates that F is in the negative direction of Fig. 13.5 when φ is in the positive direction, and vice versa. Therefore, along the arc, the acceleration $a = F/m = -g \sin \varphi$. But $\sin \varphi = x/l$ where x is the horizontal displacement of m.

$$\therefore\ a = -gx/l. \tag{13.9}$$

Now for small displacements x is approximately equal to the displacement y along the arc, since $x = l \sin \varphi$ and $y = l\varphi$; e.g. if $\varphi = 0.100$ rad $\simeq 5.7°$, $\sin \varphi = 0.09983$.

$$\therefore\ \text{for small displacements, } a = -gy/l. \tag{13.10}$$

Thus small displacements y are proportional to the accelerations a in the opposite direction, and therefore, in accordance with eqn. (13.6), the motion is simple harmonic. Moreover, from eqns. (13.6) and (13.2), $\sqrt{(-a/y)} = \omega$ and $\omega = 2\pi/T$, where T is the period of swing. But from eqn. (13.10), $-a/y = g/l$.

$$\therefore\ T = 2\pi/\omega = 2\pi\sqrt{(l/g)} \tag{13.11}$$

for small swings, independently of the particular mass.

Solution

(a) Since the bob spends 1.00 s on each side of the mid-point, $T = 2.00$ s. Then, from eqn. (13.11),

$$l = gT^2/(4\pi^2) = 9.81 \times 2.00^2/(4\pi^2) = 0.99 \text{ m.} \tag{a}$$

(b) From eqn. (13.8), $E = \frac{1}{2}m\omega^2Y^2$, where $m = 0.100$ kg, $\omega = 2\pi/T = \pi$ rad s^{-1}, and $Y = l\varphi_{max} = 0.99 \times (5\pi/180)$ m with φ_{max} in radians.

$$\therefore\ E = \tfrac{1}{2} \times 0.100 \times \pi^2 (0.99 \times 5\pi/180)^2 = 3.7 \times 10^{-3} \text{ J.} \tag{b}$$

Alternatively, E equals the difference of potential energy between the highest and lowest levels of the bob where its velocity is zero.

$$\therefore \; E = mg(l-l\cos\varphi_{max}) = 0.100\times9.81\times0.99(1-0.99619)$$
$$= 3.7\times10^{-3}\text{ J.} \qquad \text{(b)}$$

13.12. Calculate (a) the length of a simple pendulum which has a period of 3.0 s, and (b) the frequency of oscillation of another simple pendulum 25 cm in length. [(a) 2.2 m; (b) 1.0 Hz.]

13.13. A pendulum bob of mass 120 g swings on a long cord with a period of 0.50 s and amplitude of 7.5 cm. Calculate the velocity and acceleration at (a) the mid-point, (b) the extreme positions, and (c) halfway between (a) and (b). Calculate also (d) the energy of the motion. [(a) 0.94 m s^{-1}, 0; (b) 0, 12 m s^{-2}; (c) 0.82 m s^{-1}, 6.0 m s^{-2}; (d) 5.3×10^{-2} J.]

13.14. Calculate the maximum velocity of the bob of a simple pendulum which swings with an amplitude of 4.0° and period of 0.90 s. Assume standard acceleration of gravity. [0.10 m s^{-1}.]

13.15. A simple pendulum of length 25.0 cm has a period of 1.00 s. Calculate (a) the acceleration of gravity, and (b) the percentage change of period if the length is increased by 1.0 mm. [(a) 9.9 m s^{-2}; (b) 0.2%.]

13.16. A clock is controlled by a simple pendulum consisting of a thin brass rod of mass which is negligible compared with the small bob at its end. Calculate the error in seconds per day (24 h) due to expansion of the rod when the temperature rises by 10 deg C assuming $g = 9.81$ m s^{-2}. Calculate also the percentage change of the acceleration of gravity which would cause the same error. Ref.: brass. [−8.2 s d^{-1}; −0.019%.]

Vertical Elastic Motion

13.17. Worked Example

A light helical spring, which exerts a tension of 0.040 N when stretched by 1.0 mm, hangs vertically. A mass of 100 g is attached to the lower end, and is allowed to fall from rest at the position where the spring is unstrained. Calculate (a) the frequency, and (b) the displacement amplitude of the ensuing motion.

Introduction

If, as assumed, the spring is not strained beyond the elastic limit, Hooke's law applies as in exs. 6.1 and 6.2. Therefore

$$\text{tension} \propto \text{extension} \quad \text{or} \quad \text{tension} = Se,$$

where S is the constant "stiffness" of the spring and e is the extension. Then, when the body of mass m is in its steady position,

$$\text{the upward resultant force} = \text{tension} - \text{weight} = Se - mg = 0. \tag{13.12}$$

If next the body is vibrating vertically, at any instant when the body is at a height y above the steady position shown in Fig. 13.6, the tension $= S(e-y)$. Then still reckoning the upward direction to be positive, and eliminating (mg) by using eqn. (13.12),

$$\begin{aligned}\text{the resultant force} &= \text{tension} - \text{weight} \\ &= S(e-y) - mg = S(e-y) - Se = -Sy.\end{aligned}$$

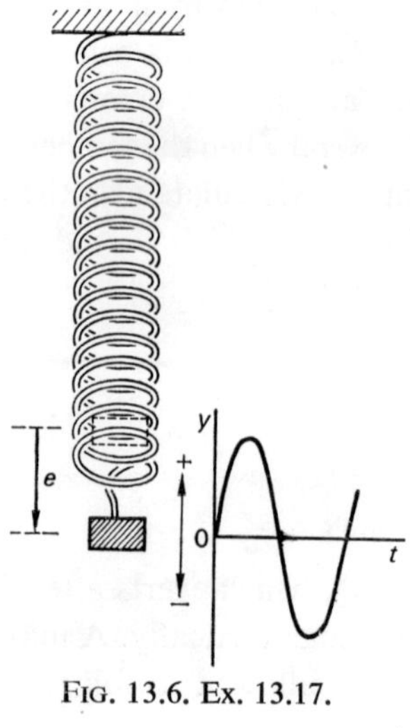

FIG. 13.6. Ex. 13.17.

This produces acceleration

$$a = -Sy/m, \qquad (13.13)$$

which shows $a \propto -y$, and therefore S.H.M. in accordance with eqn. (13.6).

$$\therefore\ \omega^2 = -a/y = S/m$$

and the period $$T = 2\pi/\omega = 2\pi\sqrt{(m/S)}. \qquad (13.14)$$

If the mass of the spring itself is appreciable, oscillation may occur even without an added mass as if the spring were light but had a mass m_0 permanently at its end. Then if m is added, the m in eqns. (13.13) and (13.14) should be replaced by the total effective mass $(m+m_0)$. Thus

$$T = 2\pi\sqrt{[(m+m_0)/S]}. \qquad (13.15)$$

It may be shown that m_0 equals one-third of the mass of the spring.

Solution

(a) Assuming negligible mass of the spring itself, eqn. (13.14) may be used with $m = 0.10$ kg, and $S = 0.040/10^{-3}$ N m^{-1}. Then

$$\text{the frequency} = 1/T = (1/2\pi)\sqrt{(S/m)}$$
$$= (1/2\pi)\sqrt{(40/0.10)} = 3.2 \text{ Hz}. \qquad \text{(a)}$$

(b) The maximum acceleration is that of gravity when the spring exerts no tension at the highest point where the motion starts. Therefore from eqn. (13.5), maximum acceleration equals $\omega^2 Y = g$.

Then the displacement amplitude $Y = g/\omega^2 = gm/S$ from eqn. (13.14).

$$\therefore\ Y = 9.81\times 0.10/40 = 0.025 \text{ m} = 2.5 \text{ cm}. \qquad \text{(b)}$$

13.18. A light helical spring hanging vertically is stretched by 0.20 cm when a load of mass 250 g is attached to the lower end.

Calculate the load required to give a vertical oscillation of frequency 5.0 Hz. [1.2 kg.]

13.19. A load of 1.0 kg hangs at the end of a light helical spring. The amplitude of vertical oscillation is 1.0 cm and the frequency is 2.0 Hz. Calculate (a) the stiffness of the spring, and (b) the energy of the motion. [(a) 160 N m^{-1}; (b) 7.9×10^{-3} J.]

13.20. A helical spring has considerable mass and has its turns sprung in contact when unloaded. When the spring is hung vertically and is loaded at the lower end, the extension is 8.0 cm for a load of 150 g and 20.0 cm for a load of 300 g. Calculate the acceleration of gravity if the time for 20 vertical oscillations is 14.8 s for the 150 g load and 20.3 s for 300 g. [9.8 m s^{-2}.]

Torsional and Compound Pendulums

13.21. Worked Example

A torsional pendulum consists of a vertical wire fixed at its upper end and supporting a brass disc in a horizontal plane at its lower end. When the disc is released after being turned to twist the wire, the pendulum undergoes 24 torsional oscillations per minute. A piece of iron is then attached to the disc so that both these pieces of metal and the wire oscillate about the axis of the disc. The frequency of oscillation is then 20 per minute. If the moment of inertia of the disc is 3.5×10^{-3} kg m^2, calculate (a) the torsional constant of the wire, and (b) the moment of inertia of the iron about the axis of oscillation.

Introduction

The pendulum shown in Fig. 13.7 oscillates when released after twisting because the torque, due to torsion of the wire, acts against angular displacement in either direction, and the momentum of the moving disc causes it to overshoot the mean position whenever it returns. The angular displacement θ at time t is indicated by the displacement of the radius OP from the stationary radial direction OA representing the mean position. The torque is proportional to θ as shown by eqn. (6.4), and may be written as $-c\theta$, where c is the "torsion constant" and the negative sign shows the direction to be opposite to that of θ. Then, if I is the moment of inertia of the disc and that of the wire is negligible, eqn. (12.1) gives angular acceleration

$$\alpha = -c\theta/I. \tag{13.16}$$

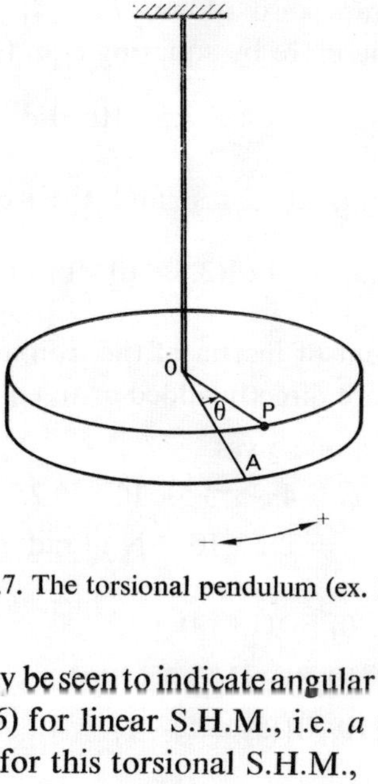

FIG. 13.7. The torsional pendulum (ex. 13.21).

This equation may be seen to indicate angular S.H.M. by comparison with eqn. (13.6) for linear S.H.M., i.e. $a = -\omega^2 y$. The comparison shows that, for this torsional S.H.M.,

$$\omega^2 = c/I. \tag{13.17}$$

Alternatively, eqn. (13.16) may be written as $d^2\theta/dt^2 = -c\theta/I$, and then a simple harmonic solution may be shown to fit by differentiation of the solution $\theta = \theta_{max} \sin \omega t$. Thus, $d\theta/dt = \omega\theta_{max} \cos \omega t$, $d^2\theta/dt^2 = -\omega^2\theta_{max} \sin \omega t = -\omega^2\theta$, and therefore eqn. (13.17) follows.

Using eqn. (13.17), the period of oscillation

$$T = 2\pi/\omega = 2\pi\sqrt{(I/c)}. \tag{13.18}$$

Finally, note that the constant angular frequency ω is not to be confused with the varying angular velocity of the disc, i.e. the constant rotation of the auxiliary circle is not the same as the simple harmonic rotation of the pendulum.

Solution

(a) With the brass disc alone, $T_1 = \frac{60}{24} = 2.5$ s, and $I_1 = 3.5\times10^{-3}$ kg m². Therefore by squaring eqn. (13.18),

$$(2.5)^2 = 4\pi^2(3.5\times10^{-3})/c. \tag{13.19}$$

Similarly, with both the brass and the iron, $T_2 = \frac{60}{20} = 3.0$ s, and so

$$(3.0)^2 = 4\pi^2[(3.5\times10^{-3})+I]/c, \tag{13.20}$$

where I is the moment of inertia of the iron and the two separate moments of inertia are directly added as in eqn. (12.6).

Equation (13.19) alone gives

$$c = 4\pi^2\times3.5\times10^{-3}/6.25$$
$$= 2.2\times10^{-2} \text{ N m rad}^{-1}. \tag{a}$$

(b) Dividing (13.20) by (13.19) gives

$$9/6.25 = [(3.5\times10^{-3})+I]/(3.5\times10^{-3}).$$
$$\therefore\ I = (3.5\times10^{-3})((9/6.25)-1) = 1.5\times10^{-3} \text{ kg m}^2. \tag{b}$$

13.22. Worked Example

A thin uniform rod of length 1.0 m swings with an amplitude of 5.0° about a horizontal axis at one end. Calculate (a) the frequency of oscillation, and (b) the maximum linear velocity of the centre of gravity.

Introduction

A compound pendulum is like a simple pendulum in that weight produces a restoring action against displacement in either direction and that momentum causes the pendulum to overshoot its mean position on each return. However, since various parts of the mass of a compound pendulum are distributed at various radii from the axis, these parts have different displacements, velocities,

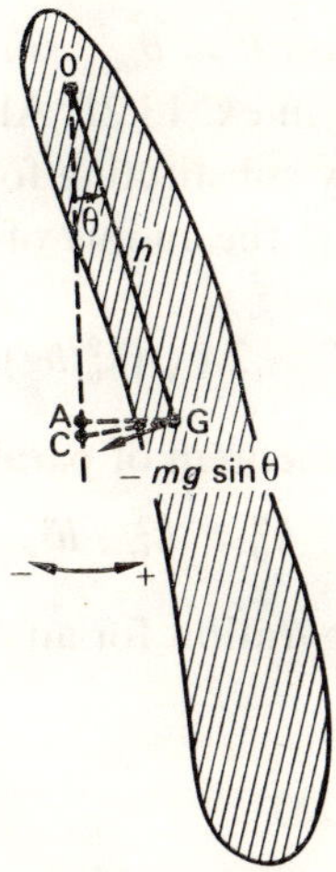

FIG. 13.8. The compound pendulum (ex. 13.22).

and accelerations. The resultant motion then depends upon the total moment of inertia as in a torsional pendulum (ex. 13.21).

Figure 13.8 shows the pendulum displaced about an axis through O perpendicular to the plane of the diagram. The unopposed component of the weight gives a turning moment of $-hmg \sin \theta$, where h is the distance of the centre of gravity G from the axis, and the negative sign is in accordance with the sign convention shown on the diagram. Then the resulting angular acceleration

$$\alpha = -hmg \sin \theta / I, \qquad (13.21)$$

where I is the moment of inertia about the axis through O.

Now, as for a simple pendulum (ex. 13.11), if θ is small, $\sin\theta = \theta$ radians to a good approximation, since $AG = h \sin\theta$ and arc $CG = h\theta$. Then eqn. (13.21) becomes

$$\alpha = -hmg\theta/I \tag{13.22}$$

which shows $\alpha \propto -\theta$ and therefore indicates angular S.H.M., with period

$$T = 2\pi/\omega = 2\pi\sqrt{[I/(hmg)]}. \tag{13.23}$$

Alternatively, a solution $\theta = \theta_{max} \sin\omega t$ may be shown to fit by finding $\alpha = d^2\theta/dt^2$, as in ex. 13.21. Also T or ω may be shown to be independent of m by substituting for I in eqn. (13.23). Thus, since $I = mk_0^2$, where k_0 is the radius of gyration about the axis through O,

$$T = 2\pi\sqrt{(k_0^2/hg)}. \tag{13.24}$$

Finally, note that the theorem of parallel axes gives

$$k_0^2 = k_G^2 + h^2 \tag{13.25}$$

where k_G is the radius of gyration for an axis through G parallel to that through O.

Solution

(a) From eqn. (12.13) for a uniform rod or bar of negligible thickness, $I_G = mk_G^2 = ml^2/12$, where the length of the rod $l = 1.0$ m.

$$\therefore\ k_G^2 = 1.0/12 = 0.083 \text{ m}^2.$$

Then, using eqns. (3.24) and (13.25), the frequency

$$f = 1/T = (1/2\pi)\sqrt{[(gh)/(h^2+k_G^2)]}$$

$$= 0.159[9.81\times0.50/(0.25+0.083)]^{\frac{1}{2}} = 0.61 \text{ Hz.} \quad \text{(a)}$$

(b) The linear amplitude of G equals the length of arc CG for a radius $h = 0.50$ m and an angle $\theta = 5.0\times\pi/180$ rad.

$$\therefore\ \text{amplitude } Y = 0.50\times(5.0\times\pi/180) = 0.0437 \text{ m.}$$

Then the maximum linear velocity $V = \omega Y$, where $\omega = 2\pi f = 2\pi \times 0.61$.

$$\therefore \; V = 2\pi \times 0.61 \times 0.0437 = 0.17 \text{ m s}^{-1}. \qquad \text{(b)}$$

13.23. The period of oscillation of a body of mass 0.50 kg, swinging as a compound pendulum, is the same as that of a simple pendulum of length 0.50 m. Calculate the moment of inertia of the body about its axis of suspension if this axis is 0.20 m from the centre of gravity. [0.050 kg m^2.]

13.24. A torsional pendulum oscillates with a period of π seconds and angular displacement amplitude of 0.10 rad. Calculate (a) the angular frequency, (b) the angular velocity amplitude, and (c) the time for the pendulum to complete the last 0.010 rad of displacement before reaching its full amplitude in either direction. [(a) 2.0 rad s^{-1}; (b) 0.20 rad s^{-1}; (c) 0.23 s.]

13.25. A torsional pendulum consists of a horizontal disc having a moment of inertia of 4.0×10^{-3} kg m^2 about its normal axis of symmetry, suspended on a vertical wire of which the torsion constant is 0.30 Nm rad^{-1}. The displacement amplitude of torsional oscillation is 30°. Calculate the angular velocity of the oscillation (a) when the displacement is half the maximum value, and (b) when the kinetic energy is half its maximum value. [(a) 3.9 rad s^{-1}; (b) 3.2 rad s^{-1}.]

13.26. A uniform bar, of length 1.00 m and square cross-section of side 2.00 cm, swings about an axis through the bar 30 cm from one end, perpendicular to two sides and equidistant from the other sides. (a) Calculate the period of oscillation, assuming the cross-section dimensions to be negligible compared with the length. (b) Calculate the error due to the above assumption. [(a) 1.57 s; (b) -0.0135%.]

CHAPTER 14

TRANSFER OF HEAT

Thermal Conduction

14.1. WORKED EXAMPLE

A rectangular sheet of glass, $80 \times 50 \times 0.40$ cm, has air of main temperatures $-5°C$ and $+25°C$ on its two sides. Calculate the rate of flow of heat through the sheet (a) assuming that the faces of the glass have the above temperatures, and (b) assuming that the air with these temperatures is separated from the glass by a stationary air film in contact with each face and of effective thickness 1.0 mm. Ref.: air and glass.

Introduction

Heat conduction is the transfer of heat from one part of a substance to another in contact with it by a collision process in which the faster moving particles of the hotter region transfer some of their thermal kinetic energy to the slower particles of the colder region. Conduction always occurs from hotter to colder regions, even when these are of different substances.

When heat is conducted through a material, it is found that the rate of flow of heat is proportional to the cross-section area and to the temperature gradient. Thus

$$Q/t = \lambda A(\delta\theta/\delta x), \tag{14.1}$$

where Q is the heat which flows in a time t through a cross-section area A; $\delta\theta/\delta x$ is the temperature gradient, i.e. $\delta\theta$ is the fall of temperature for a distance δx in the direction of flow, and λ is a constant of proportionality called the thermal conductivity for the particular material.

The units of λ are found by rearranging eqn. (14.1) and substituting the appropriate units for the other symbols of the equation, cancelling if possible. Thus

$$\lambda = \frac{Q/t}{A\delta\theta/\delta x},$$

and with M.K.S. units,

$$\frac{\text{J/s}}{\text{m}^2 \text{ deg C/m}} = \text{W m}^{-1} \text{deg C}^{-1}.$$

If heat flows along a bar with well-lagged sides, so that heat loss from the sides is negligible, the rate of heat flow and the temperature gradient may be considered uniform. Similarly, uniformity is assumed when heat flows through a sheet of material, such as a window pane, when losses from the edges may be neglected because of their small surface areas. Then, if the temperatures θ_1 and θ_2 are uniform over the two faces, a distance x apart, $\delta\theta/\delta x = (\theta_1 - \theta_2)/x$.

If heat flows through one substance and then through another of different conductivity, the temperature gradients will be different for equal rates of heat flow through the two materials. This applies to heat flowing through a window pane, since wind and convection of the air on either side are opposed by the adhesion of air to the glass faces. The effect is as if each face of the glass were covered by a stationary film of air of some effective thickness through which conduction occurs without convection or wind. A considerable drop of temperature may occur through each film, so reducing the drop through the glass to much less than the difference of the main temperatures of the moving air inside and outside the window. Thus heat loss through the window depends considerably upon these surface films.

Solution

(a) Area $A = 80 \times 50 \times 10^{-4} = 0.40\ \text{m}^2$.

Neglecting air films $\delta\theta/\delta x = (\theta_1 - \theta_2)/x = 30/(4.0 \times 10^{-3})$
$= 7.5 \times 10^3$ deg C m^{-1}.

From the reference table λ for glass $= 0.80$ W m^{-1} deg C^{-1}. Then eqn. (14.1) gives

$$Q/t = \lambda A \delta\theta/\delta x = 0.80 \times 0.40 \times 7.5 \times 10^3$$
$$= 2.4 \times 10^3 \text{ W} = 2.4 \text{ kW}. \qquad \text{(a)}$$

(b) Now consider the air films for which $\lambda = 0.023$ W m^{-1} deg C^{-1}. If the temperatures are as in Fig. 14.1, applying eqn.

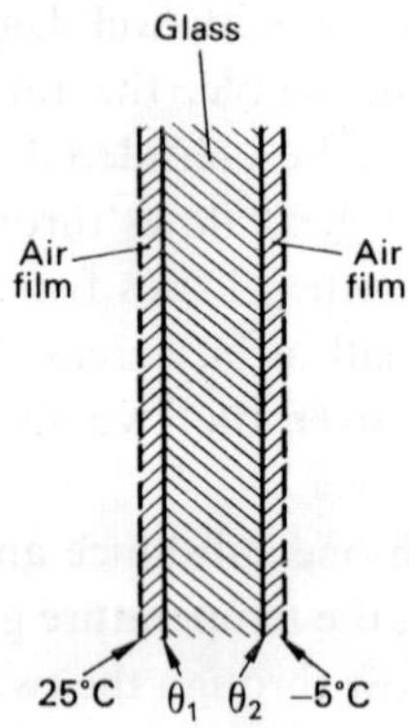

FIG. 14.1. Thermal conduction (ex. 14.1).

(14.1) to the warmer air film gives

$$Q/t = 0.023 \times 0.40 \times (25 - \theta_1)/(1.0 \times 10^{-3}).$$

Similarly, for the colder film,

$$Q/t = 0.023 \times 0.40 \times (\theta_2 + 5)/(1.0 \times 10^{-3}) = 9.2(\theta_2 + 5),$$

where Q/t is the same for both films and glass. Then

$$25 - \theta_1 = \theta_2 + 5, \quad \therefore\ \theta_1 + \theta_2 = 20°\text{C}.$$

For the glass $$Q/t = 0.80\times 0.40\times(\theta_1-\theta_2)/(4.0\times 10^{-3})$$
$$= 80(\theta_1-\theta_2).$$

But $\theta_1+\theta_2 = 20$. Therefore $Q/t = 80(20-2\theta_2) = 9.2(\theta_2+5)$ from above

$$\therefore\ 1600-160\theta_2 = 9.2\theta_2+46.$$
$$\therefore\ \theta_2 = 1554/169.2 = 9.2^\circ\text{C}.$$
$$\therefore\ Q/t = 9.2\times(9.2+5)$$
$$= 131\text{ W} = 0.13\text{ kW}. \qquad \text{(b)}$$

14.2. Worked Example

An asbestos tube having inner and outer diameters of 2.0 cm and 4.0 cm surrounds a heating element which dissipates a power of 5.0 W uniformly along a length of 25 cm. The outer surface of the tube is maintained at a temperature of 100°C. Calculate the temperature of the inner surface. Ref.: asbestos.

Introduction

Since the areas A_1 and A_2 of the inner and outer curved surfaces of a tube are different, the temperature gradient is not constant for heat flowing through the wall. Equation (14.1) may be modified to give the negative value of the temperature gradient in the direction of the radius r at each surface, inner 1 and outer 2. Thus

$$(d\theta/dr)_1 = -(Q/t)/(\lambda A_1)$$

and

$$(d\theta/dr)_2 = -(Q/t)/(\lambda A_2),$$

where Q/t is the rate of flow of heat, λ is the thermal conductivity, and the negative sign is necessary since θ decreases as r increases.

An approximate calculation of the heat flow may be made if the wall is thin relative to the radii r_1 and r_2. For this, the mean area

$$A = (\text{length of tube})\times(\text{mean circumference}) = l\pi(r_1+r_2) \qquad (14.2)$$

and

$$\text{the mean temperature gradient} = \delta\theta/(r_2 - r_1), \tag{14.3}$$

where $\delta\theta$ is the temperature difference of the inner and outer surfaces, and $(r_2 - r_1)$ is the wall thickness.

The error in using these mean values may be acceptable when $r_2/r_1 < 2$, but increases rapidly as the ratio of radii increases.

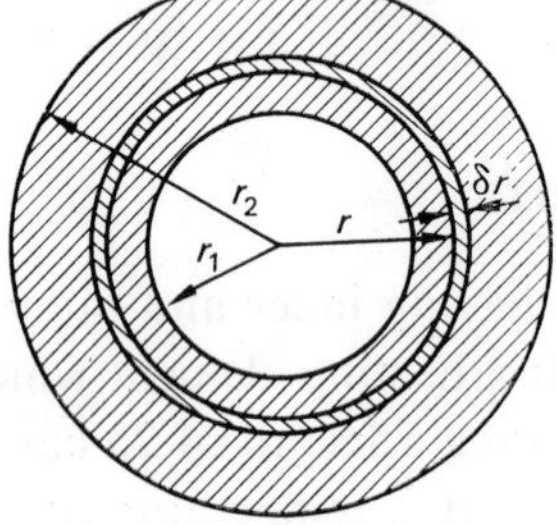

FIG. 14.2. Elementary part of thick tube.

An exact calculation for any radii is given by integration, starting with an elementary tube of infinitesimal thickness, as in Fig. 14.2.

Then eqn. (14.1) is modified to give $Q/t = -\lambda(2\pi rl)\, d\theta/dr$ for the element. Then, for the complete tube,

$$\frac{\lambda 2\pi l}{Q/t} \int_{\theta_1}^{\theta_2} -d\theta = \int_{r_1}^{r_2} \frac{dr}{r}.$$

$$\therefore \quad \frac{\lambda 2\pi l}{Q/t}(\theta_1 - \theta_2) = \log_e (r_2/r_1). \tag{14.4}$$

Solution

First by the approximate method: Equation (14.2) gives the mean area through which heat flows,

$$\begin{aligned} A_{\text{mean}} &= l\pi(r_1 + r_2) = 0.25\pi(0.01 + 0.02) \\ &= 7.5\pi \times 10^{-3}\ \text{m}^2. \end{aligned}$$

Equation (14.3) gives the mean positive value of the temperature gradient,

$$\delta\theta/(r_2-r_1) = (100-\theta_2)/(0.02-0.01)$$
$$= (100-\theta_2)100 \text{ deg C/m}.$$

Then, from eqn. (14.1) and using $\lambda = 0.15$ for the conductivity of asbestos,

$$Q/t = \lambda A_{\text{mean}}\delta\theta/(r_2-r_1) = 0.15(7.5\pi\times10^{-3})(100-\theta_2)100$$
$$= 0.354(100-\theta_2) = 5.0 \text{ W},$$

stated in the data.

$$\therefore\ \theta_2 = (35.4-5.0)/0.354 = 86°\text{C}.$$

Secondly, using the logarithmic eqn. (14.4) for a more accurate result:

$$\frac{0.15\times2\pi\times0.25}{5.0}(100-\theta_2) = \log_e(0.02/0.01),$$

$$\therefore\ 0.0471(100-\theta_2) = 0.693.$$

$$\therefore\ \theta_2 = (4.71-0.693)/0.0471 = 85°\text{C}.$$

14.3. An oven in the form of a cube with edges of 50 cm has walls insulated with asbestos 1.0 cm thick. Calculate the power of an internal heater required to maintain a temperature difference of 120 deg C between the inner and outer surfaces of the asbestos. Ref.: asbestos. [2.7 kW.]

14.4. Calculate the rate of loss of heat through a glass window $1.0\times2.0\ \text{m}^2$ in area and 5.0 mm in thickness, when the inside and outside surfaces are at temperatures of 20° and −10°C respectively. If the room temperature can be maintained solely by a heater of 1.0 kW, and if 60% of the heat is assumed to be lost through this window, calculate the maximum temperature difference which can exist between the inside and outside surfaces of the glass. Ref.: glass. [9.6 kW; 1.9 deg C.]

14.5. An aluminium plate 1.0 cm thick and a copper plate 2.0 cm thick are face to face in good thermal contract. The exposed face of the aluminium is maintained at 0°C and that of the copper at 10°C. Calculate (a) the temperature of the faces in contact, and (b) the rate of flow of heat per unit area across these faces. Ref.: aluminium and copper. [(a) 4.9°C; (b) 9.8 W cm^{-2}.]

14.6. A thin copper pipe having a length of 10 m and diameter of 5.0 cm is lagged with a layer of asbestos 1.0 cm thick. When steam at 100°C is supplied through the pipe, the outer surface of the asbestos has a temperature of 25°C. Calculate approximately the mass of steam condensed per hour. Ref.: asbestos and steam. [3 kg.]

Newton's Law of Cooling

14.7. Worked Example

A body of thermal capacity 600 J deg C^{-1} cools at a rate of 5.0 deg C min^{-1} when its temperature is 65°C, and at 3.0 deg C min^{-1} when 45°C. Assuming Newton's law of cooling to apply, calculate (a) the time constant, (b) the rate of loss of heat when the temperature is 40°C, and (c) the approximate time taken to cool from 45°C to 40°C.

Introduction

When an object at one temperature is in surroundings at another, it tends to gain or lose heat by convection, conduction, and radiation, the flow of heat being always from higher to lower temperature. When convection is forced and radiation is slight, Newton's law of cooling applies:

The rate of loss of heat from a body, at any instant, is proportional to the excess of the temperature of the body over the temperature of the surroundings.

Thus $$P \propto \Delta\theta, \qquad (14.5)$$

where P is the rate of loss of heat = power of heat loss (watts), $\Delta\theta$ is the temperature excess $= \theta - \theta_a$, where θ is the temperature of body and θ_a is the ambient temperature = temperature of surroundings.

This law does not apply accurately when the temperature excess is large, for then the loss by radiation is large and not even approximately proportional to the excess. Also if convection is not forced, the ambient temperature becomes uncertain due to irregular heating of the surrounding air or other matter.

Note that eqn. (14.5) refers to the rate of loss of heat and not to the rate of fall of temperature. Yet if the thermal capacity of the body is constant and no heat or other energy is supplied, the rate of loss of heat P is proportional to the rate of fall of temperature $d\theta/dt$, so that, with eqn. (14.5),

$$d\theta/dt \propto \Delta\theta. \tag{14.6}$$

However, when heat or other energy is supplied and lost at the same time, as in an electric heater, $d\theta/dt$ may be zero although P still follows eqn. (14.5).

Figure 14.3 shows the variation of temperature with time when an object cools in accordance with the law, the thermal capacity being constant and no energy being supplied to the object. At any point X, the negative slope of the tangent represents $-d\theta/dt$, and the height of the small triangle represents $\Delta\theta$ relative to a constant ambient temperature θ_a. Now

$$-d\theta/dt = \Delta\theta/\tau, \tag{14.7}$$

where τ is shown by the base of the triangle. But from eqn. (14.6), $-d\theta/dt \propto \Delta\theta$. Therefore τ is a constant, called the time constant. Thus the tangent at any point of the curve leads to the same value of τ, as seen in the two examples of Fig. 14.3.

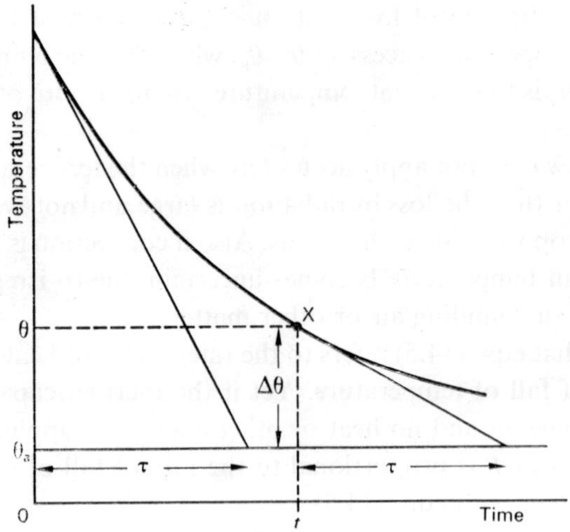

FIG. 14.3. Cooling time constant.

Solution

(a) Substituting the data in eqn. (14.7) gives

$$5 = (65-\theta_a)/\tau \quad \text{and} \quad 3 = (45-\theta_a)/\tau.$$

$$\therefore \ \theta_a = 65-5\tau = 45-3\tau.$$

$$\therefore \ \tau = 10 \text{ min.} \qquad \text{(a)}$$

Also $\theta_a = 65-5\tau = 65-50 = 15°C.$

(b) Then for $\theta = 40°C$, with $\tau = 10$ min, and $\theta_a = 15°C$, eqn. (14.7) gives

$$-d\theta/dt = (40-15)/10 = 2.5 \text{ deg C min}^{-1}.$$

But (rate of loss of heat) = (thermal capacity)×(rate of fall of temperature).

$$\therefore \ P = C(-d\theta/dt), \quad \text{where} \quad C = 600 \text{ J deg C}^{-1}.$$

$$\therefore \ P = 600\times2.5 = 1500 \text{ J min}^{-1} = 25 \text{ J s}^{-1} = 25 \text{ W}. \qquad \text{(b)}$$

(c) For small changes of θ and t the cooling curve may be taken as approximately straight, with a gradient equal to the tangent at the mid-point of that part of the curve. Thus, at 42.5°C,

$$-d\theta/dt = (42.5-15)/10 = 2.75 \text{ deg C min}^{-1}.$$

Then the approximate time is that for cooling by 5.0 deg C at this rate; i.e.

$$5.0/2.75 = 1.8 \text{ min.} \qquad \text{(c)}$$

14.8. Worked Example

An object cools from 60°C to 51°C in 2.0 min with a constant ambient temperature of 10°C maintained by forced convection. Calculate the time taken to cool from 60°C to 12.5°C under the same conditions, assuming constant thermal capacity.

Introduction

As in eqn. (14.7) the rate of fall of temperature

$$-d\theta/dt = \Delta\theta/\tau, \qquad (14.8)$$

where τ is the time constant and the excess temperature $\Delta\theta = \theta-\theta_a$, θ_a being the constant ambient temperature. This equation may be shown to have an exponential solution of the form

$$\Delta\theta = \Delta\theta_0 e^{-t/\tau} \qquad (14.9)$$

where $\Delta\theta_0$ is the excess temperature when $t = 0$ and e is the base of natural logarithms = 2.718.... Thus by differentiating eqn. (14.9),

$$d(\Delta\theta)/dt = -(1/\tau)\Delta\theta_0 e^{-t/\tau} = -\Delta\theta/\tau.$$

But $$d(\Delta\theta)/dt = d(\theta-\theta_a)/dt = d\theta/dt.$$

$$\therefore\ d\theta/dt = -\Delta\theta/\tau,$$

which agrees with eqn. (14.8).

Alternatively, eqn. (14.9) may be expressed logarithmically:

$$\log_e(\Delta\theta_0/\Delta\theta) = t/\tau. \qquad (14.10)$$

Equation (14.9) is plotted in Fig. 14.4 where the points X_1, X_2, X_3, etc., are shown for equal intervals of time t_1. Now

$$\Delta\theta_2/\Delta\theta_1 = (\Delta\theta_0 e^{-2t_1/\tau})/(\Delta\theta_0 e^{-t_1/\tau}) = e^{-t_1/\tau}.$$

Similarly,

$$\Delta\theta_3/\Delta\theta_2 = \Delta\theta_4/\Delta\theta_3 = e^{-t_1/\tau}.$$

Thus the values of $\Delta\theta$ at regular intervals of time t_1 fall as a geometric progression having a common ratio of

$$e^{-t_1/\tau}. \tag{14.11}$$

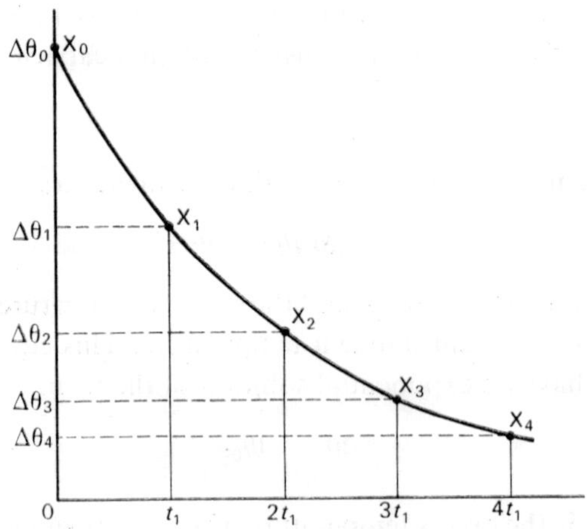

FIG. 14.4. Exponential curve and geometric progression.

Solution

(a) Using the principle of the geometric progression for n terms beyond the first, $\Delta\theta_0 = 60-10 = 50$ deg C, $\Delta\theta_1 = 51-10 = 41$ deg C after a time $t_1 = 2.0$ min, and $\Delta\theta_n = 12.5-10 = 2.5$ deg C after the full time $nt_1 = 2.0\,n$ min. Then

$$\Delta\theta_n/\Delta\theta_0 = (\Delta\theta_1/\Delta\theta_0)^n.$$

$$\therefore \log_{10} \Delta\theta_n - \log_{10} \Delta\theta_0 = n(\log_{10} \Delta\theta_1 - \log_{10} \Delta\theta_0).$$

$$\therefore n = (\log_{10} 2.5 - \log_{10} 50)/(\log_{10} 41 - \log_{10} 50)$$
$$= 1.301/0.086 = 15.$$

$$\therefore \text{time required} = nt_1 = 15 \times 2.0 = 30 \text{ min.} \qquad \text{(a)}$$

(b) Alternatively, using the exponential solution (14.9),

$$41 = 50e^{-2/\tau} \quad \text{and} \quad 2.5 = 50e^{-t/\tau},$$

where t is the required time in minutes.

From the first of these,

$$e^{-1/\tau} = \sqrt{(41/50)} = 0.906.$$

$$\therefore 2.5/50 = e^{-t/\tau} = (0.906)^t.$$

$$\therefore t = \log_{10} 0.05/\log_{10} 0.906 = 1.301/0.043 = 30 \text{ min.} \qquad \text{(b)}$$

14.9. An object of constant thermal capacity cools at a rate of 4.0 deg C min^{-1} when its temperature is 63°C, and at 1.0 deg C min^{-1} when 27°C. Assuming Newton's law of cooling, calculate (a) the time constant, (b) the ambient temperature, and (c) the rate of cooling at 51°C. [(a) 12 min; (b) 15°C; (c) 3.0 deg C min^{-1}.]

14.10. An aluminium calorimeter of mass 30 g contains 100 g of water. The rate of cooling is 2.5 deg C min^{-1} in a current of air of temperature 50 deg C below that of the calorimeter. Estimate the rate of loss of heat when a fall of 10 deg C occurs in the temperature of (a) the calorimeter alone, (b) the air current alone, and (c) both the calorimeter and the air current. Ref.: aluminium and water. [(a) 15 W; (b) 22 W; (c) 19 W.]

14.11. An empty copper can of mass 500 g cools at a rate of 50 W when its temperature is 60°C in surroundings of 15°C. Calculate the rate of fall of temperature when the can contains 400 g of water at 90°C, assuming Newton's law of cooling and a constant ambient temperature. Ref.: water and copper. [0.045 deg C s^{-1}.]

14.12. A tank of water is maintained at a temperature of 40°C in surroundings of 15°C by supplying a power of 50 W by an immersion heater. Calculate (a) the steady temperature attained if the power is doubled, and (b) the net rate of gain of heat at 85°C if the power is increased tenfold. [(a) 65°C; (b) 360 W.]

14.13. A body cools in a draught of air at a constant temperature from 45.4°C to 33.4°C in 5.0 min, and then to 26.2°C in the next 5.0 min. Calculate (a) the temperature after a third period of 5.0 min, and (b) the time taken to cool from 45.4°C to 17.5°C. [(a) 21.8°C; (b) 25 min.]

14.14. A bar of metal at an initial temperature 40.0 deg C in excess of its surroundings cools at an instantaneous rate of 5.00 deg C min^{-1}. Assuming Newton's law of cooling, calculate (a) the time taken for the temperature to fall from the initial value by 5.0 deg C, and (b) the fall of temperature which occurs in the first minute. [(a) 1.07 min; (b) 4.7 deg C.]

14.15. A piece of metal in a draught of air of temperature 20°C cools from 100°C to 60°C in a time of 10 min. Calculate (a) the time taken for the first half of the above fall of temperature, and (b) the fall of temperature during the first half of the above time interval. [(a) 4.1 min; (b) 23.4 deg C.]

Cooling Correction

14.16. Worked Example

When a sample of fuel was tested in a bomb calorimeter the temperatures given below were recorded at the stated times after the instant of igniting the fuel. Estimate the maximum temperature which would have been reached if no heat had been lost.

Time (min)	0	0.5	1.0	1.5	2.0	2.5	3.0	3.5
Temp. (°C)	15.0	15.2	15.7	16.6	17.6	18.5	19.5	20.5
Time (min)	4.0	4.5	5.0	6.0	7.0	8.0	9.0	10.0
Temp. (°C)	21.1	21.3	21.3	21.2	21.0	20.8	20.6	20.4

Introduction

When a sample of fuel is ignited in a bomb calorimeter to determine the heat it produces, some allowance may be necessary for loss of heat during the test. A correction to be added to the maximum temperature observed may be deduced as follows from the rate of cooling beyond the maximum.

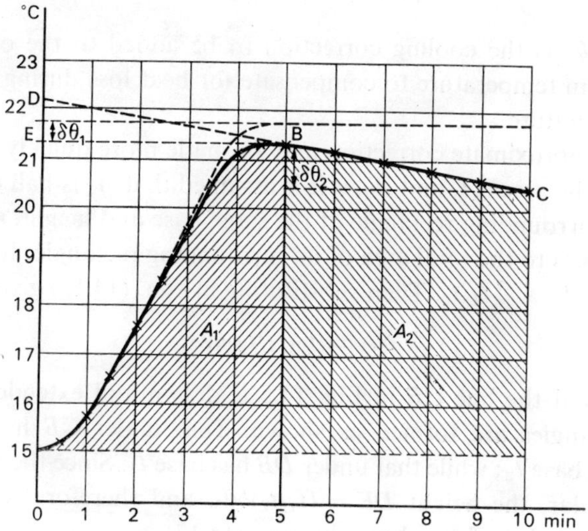

FIG. 14.5. Cooling correction (ex. 14.16).

Figure 14.5 shows the temperature plotted against time with two areas beneath the curve divided at the time of maximum temperature. Since each area equals its base length multiplied by its average height, area $A_1 = t_1(\text{average } \Delta\theta)_1$ and area $A_2 = t_2(\text{average } \Delta\theta)_2$, where t = time, suffix 1 refers to rising temperatures, suffix 2 to falling temperatures, and $\Delta\theta$ is the excess of temperature above the initial value which is assumed to be ambient level.

Now, assuming Newton's law of cooling, the proportionality of eqn. (14.5) applies. Therefore the average rate of loss of heat

$P \propto$ (average $\Delta\theta$).

$$\therefore \quad A_1 \propto P_1 t_1 \quad \text{and} \quad A_2 \propto P_2 t_2.$$

But if the thermal capacity is constant, $Pt \propto$ temperature loss during time t. Therefore during part 2 the temperature loss $\delta\theta_2 \propto A_2$, and, similarly, during part 1 the "lost rise" $\delta\theta_1 \propto A_1$.

$$\therefore \quad \delta\theta_1 = (A_1/A_2)\delta\theta_2, \tag{14.12}$$

where $\delta\theta_1$ is the cooling correction to be added to the observed maximum temperature to compensate for heat loss during the rise of temperature.

An approximate correction may be made more quickly than by finding the areas A_1 and A_2, if it is assumed that A_1 is half the area of the surrounding rectangle of the same base and height, and that A_2 is equal to the total area of its surrounding rectangle. It follows that $A_1/A_2 \simeq \frac{1}{2}t_1/t_2$. Then substituting in eqn. (14.12) gives

$$\delta\theta_1 \simeq (\tfrac{1}{2}t_1/t_2)\delta\theta_2. \tag{14.13}$$

Now if the line CB in Fig. 14.5 is drawn and extended to D, two triangles are formed as shown. That under CB has height $\delta\theta_2$ and base t_2; while that under DB has base t_1. Since the triangles are similar, the height $DE = (t_1/t_2)\delta\theta_2$, and therefore represents twice the value of $\delta\theta_1$ shown by eqn. (14.13).

Solution

The ratio of areas estimated by counting squares under the curve of Fig. 14.5 is $A_1/A_2 = 16\frac{1}{2}/29\frac{1}{2}$. Then, since $\delta\theta_2$ is seen to be 0.9 deg C, eqn. (14.12) gives

$$\delta\theta_1 = (16\tfrac{1}{2}/29\tfrac{1}{2})\times 0.9 = 0.50 \text{ deg C}.$$

The construction of the line CBD and subsequent halving of the intercept DE gives $\delta\theta_1 = 0.45$ deg C.

Due to the approximations involved in the two methods, and particularly in choosing the time at which maximum temperature would have been reached with no heat loss, it is not surprising that the two estimates of $\delta\theta_1$ agree to only one significant figure.

Then the estimated maximum temperature = 21.3+0.5 = 21.8°C.

14.17. Estimate the temperatures which would have been reached in the test of ex. 14.16 at times of 3.0 min and 4.0 min if no heat had been lost during the test. [19.7°C; 21.4°C.]

Radiation

14.18. Worked Example

A radiant heater is required to give a total radiation of 1.0 kW from a surface of temperature 2500°C. Calculate (a) the surface area required, and (b) the wavelength and frequency at which peak radiation would occur. Assume that the surface is a full black-body radiator.

Introduction

Radiation is the transmission of energy by waves of electric and magnetic fields. In contrast to conduction and convection, no material is necessary to convey energy by radiation and until the energy is absorbed in something it does not appear as heat.

Most electromagnetic radiation is invisible, but the eye responds to light in a particular band of frequencies or wavelengths which produce sensations of colour in addition to conveying energy which may become heat. At room temperatures objects produce invisible infrared radiations, and even when "red hot" their radiation is mostly infrared although it extends into the visible band.

In vacuum, and approximately in air, all electromagnetic radiations travel at the same velocity, $c = 3.00 \times 10^8$ m s^{-1}. Since the frequency f is the number of waves passing a point per unit time, and the wavelength λ is the length of each wave, $f\lambda$ is the total length of the waves passing per unit time, i.e.

$$f\lambda = \text{velocity } c. \qquad (14.14)$$

The maximum possible total power P_{max} of radiation emitted by a surface is given by the Stefan–Boltzmann law:

$$P_{max} = A\sigma T^4, \tag{14.15}$$

where P_{max} is in watts, A is the surface area in metre^2 at Kelvin temperature T, and σ is the Stefan–Boltzmann constant $= 5.67 \times 10^{-8}$ W m^{-2} K^{-4}. The actual radiation depends considerably upon the nature and condition of the surface. For example, a dull black surface may radiate almost the maximum for its temperature, but a polished silvered surface would radiate considerably less. Hence a factor ε called the emissivity of the surface is used. Thus

$$P = \varepsilon A\sigma T^4, \tag{14.16}$$

where ε is unity only for a so-called "full radiator" or "perfect black body".

Wien's displacement law states that the wavelength λ_p of the peak radiation emitted by a full radiator is inversely proportional

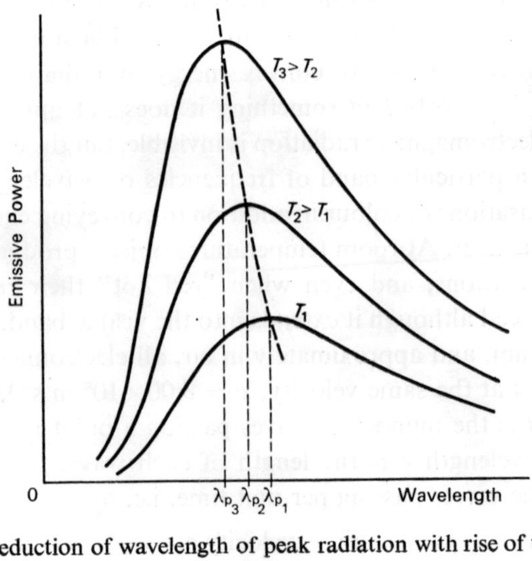

FIG. 14.6. Reduction of wavelength of peak radiation with rise of temperature.

to the Kelvin temperature T. Thus

$$\lambda_P T = 2.9\times10^{-3} \text{ m K}, \qquad (14.17)$$

where λ_P is in metres. The effect is shown in Fig. 14.6, where the peak of radiation shifts toward shorter wavelengths as the temperature rises, and the area under each curve, if completed, would represent the total power of radiation per unit area at that temperature.

Solution

(a) When $P = 10^3$ W and $T = 2500+273$ K, with $\varepsilon = 1$ for full radiation and $\sigma = 5.67\times10^{-8}$ W m^{-2} K^{-4}, eqn. (14.15) or (14.16) gives

$$A = P/(\varepsilon\sigma T^4) = 10^3/(5.67\times10^{-8}\times2773^4)$$
$$= 3.0\times10^{-4} \text{ m}^2 = 3.0 \text{ cm}^2. \qquad \text{(a)}$$

(b) From eqn. (14.17), $\lambda_p = 2.9\times10^{-3}/2773$
$$= 1.05\times10^{-6} \text{ m}. \qquad \text{(b)}$$

Then, from eqn. (14.14), $f = c/\lambda = 3.00\times10^8/(1.05\times10^{-6})$
$$= 2.9\times10^{14} \text{ Hz}. \qquad \text{(b)}$$

14.19. Worked Example

A copper can of mass 100 g has a blackened outer surface of area 150 cm^2 and contains 100 g of water. When the temperature of the can is 60°C and the ambient temperature is 10°C the rate of cooling is 2.0 deg C min^{-1}. Calculate (a) the rate of supply of heat to maintain a temperature of 60°C, and (b) the proportion of the cooling which is due to radiation if the emissivity is 0.60. Ref.: specific heats and Stefan–Boltzmann constant.

Introduction

If one body at temperature T_1 is completely surrounded by another at temperature T_2, the resultant rate at which energy is radiated outward is found by considering both radiation and absorption to occur simultaneously. Now if $T_1 = T_2$, the inner body radiates and absorbs equally, so maintaining a constant temperature if no other power but radiation is involved. So the inner body absorbs at the rate given by eqn. (14.16): $P = \varepsilon A\sigma T_2^4$. Then if this absorption continues when $T_1 > T_2$, the resultant outward radiation is given by

$$P = \varepsilon A\sigma T_1^4 - \varepsilon A\sigma T_2^4 = \varepsilon A\sigma(T_1^4 - T_2^4). \tag{14.18}$$

If the inner body has a power supply, as in the case of an electric heater, a constant temperature is maintained when the power supplied equals the power lost by radiation, conduction and convection.

Solution

(a) The gain or loss of heat for a given rise or fall of temperature depends upon the thermal capacity as in ex. 14.7. Here the thermal capacity $= (mc)_{\text{copper}} + (mc)_{\text{water}}$, where for the copper $m = 0.100$ kg and $c = 390$ J kg^{-1} deg C^{-1}, and for the water $m = 0.100$ kg and $c = 4190$ J kg^{-1} deg C^{-1}.

Then the power required = rate of loss of heat

$$\begin{aligned} &= (\text{thermal capacity}) \times (\text{rate of fall of temperature}) \\ &= [(0.100 \times 390) + (0.100 \times 4190)] \times 2.0/60 \text{ W} \\ &= 0.100 \times 4580 \times 2.0/60 = 15 \text{ W}. \end{aligned} \tag{a}$$

(b) The radiation loss is given by eqn. (14.18), in which $\varepsilon = 0.60$, $A = 150 \times 10^{-4}$ m^2, $\sigma = 5.67 \times 10^{-8}$ W m^{-2} K^{-4}, $T_1 = 273 + 60$ K, and $T_2 = 273 + 10$ K.

$$\begin{aligned} \therefore \text{ radiant power} &= 0.60 \times 150 \times 10^{-4} \times 5.67 \times 10^{-8}(333^4 - 283^4) \\ &= 5.1 \times 10^{-10} \times (1.23 - 0.64) \times 10^{10} = 3 \text{ W}. \end{aligned}$$

Then 3 W as a proportion of the total 15 W is one-fifth. (b)

14.20. The tungsten filament of a lamp has a surface area of 1.5 cm^2. Calculate the temperature maintained by a power supply of 150 W assuming an emissivity of 40%, and all heat loss to be by radiation. Ref.: Stefan–Boltzmann constant. [2600 K.]

14.21. A block of steel of mass 50 g and surface area 20 cm^2 cools in surroundings maintained at a temperature of 20°C. Assuming an emissivity of 0.50, calculate the rate of fall of temperature at 500°C. Ref.: Stefan–Boltzmann constant and steel. [0.87 deg C s^{-1}.]

14.22. A satellite in free space receives solar radiation of intensity 1.4 kW m^{-2}. Calculate the equilibrium temperature of the satellite assuming its surface to be spherical and of material which is a perfect conductor of heat and a full radiator. Assume that the satellite has no internal power supply. Ref.: Stefan–Boltzmann constant. [7°C.]

14.23. Assuming the sun to be a full radiator, calculate from the following data (a) the surface temperature of the sun, and (b) the wavelength of its peak radiation. The radius of the sun is 7.0×10^8 m and its mean distance from the earth is 1.5×10^{11} m. The intensity of solar radiation received by the earth is 1.4 kW m^{-2}. Ref.: Stefan–Boltzmann and Wien's constants. [(a) 5.8×10^3 K; (b) 0.50 μm.]

14.24. A lamp filament runs at a temperature of 2000 K when it draws a power of 32 W. Calculate the power required to run at 2500 K and the wavelength of peak radiation at this temperature. Calculate also the power and temperature when the voltage is 10% less than that for 2500 K, assuming resistance to be proportional to absolute temperature. Assume full radiation. Ref.: Stefan–Boltzmann and Wien's constants. [78 W; 1.2 μm; 66 W; 2400 K.]

CHAPTER 15

MOLECULAR THEORY OF GASES AND VAPOURS

Avogadro's Law

15.1. WORKED EXAMPLE

Estimate the mass of neon which occupies a volume of 10 cm^3 at a temperature of 20°C and a pressure of 800 mmHg. Assume neon to be an ideal gas of relative molecular mass 20.2, and assume any data of the reference section.

Introduction

Two alternative solutions (a) and (b) are given, both based on Avogadro's law, that equal volumes V of all gases at the same temperature T and pressure p contain equal numbers N of molecules.

Now if M_r is the relative molecular mass, often called molecular weight, of a particular gas, the mass of one molecule m_1 is given by $M_r m_c/12$, where m_c is the mass of a single atom of carbon having the standard atomic mass of 12.

Then the density

$$\varrho = Nm_1/V = (N/V)M_r(m_c/12). \tag{15.1}$$

But, according to Avogadro's law, N/V is the same for all gases at the same p and T.

$$\therefore\ \varrho \propto M_r \tag{15.2}$$

under the same conditions.

It follows that if a volume of gas contains a mass numerically equal to its relative molecular mass, making $\varrho = M_r/V$, V must be a constant to satisfy (15.2). This volume is 22.4 m^3 at s.t.p. for a mass M_r kilograms, i.e. for one kilomole of gas.

The effect of varying p and T is given in eqn. (9.11):

$$\varrho_2 = \varrho_1(T_1/T_2)(p_2/p_1) \tag{15.3}$$

Therefore $\varrho \propto p/T$ and, including (15.2), $\varrho \propto M_r p/T$.

$$\therefore\ M_r p/T = R\varrho, \tag{15.4}$$

where R is the universal gas constant $= 8.31 \times 10^3$ in M.K.S. units, namely J K^{-1} $kmole^{-1}$.

Further, if m is the total mass of gas, $\varrho = m/V$, and eqn. (15.4) gives

$$pV/T = Rm/M_r. \tag{15.5}$$

Solution

(a) In M.K.S. units $p = 800 \times 10^{-3} \times 13.6 \times 10^3 \times 9.81 = 1.07 \times 10^5$ N m^{-2} and $V = 10 \times 10^{-6} = 10^{-5}$ m^3. Also $T = 293$ K, and $M_r = 20.2$. Therefore using eqn. (15.5), the mass of neon

$$\begin{aligned} m &= pVM_r/RT \\ &= 1.07 \times 10^5 \times 10^{-5} \times 20.2/(8.31 \times 10^3 \times 293) \\ &= 8.8 \times 10^{-6} \text{ kg} = 8.8 \text{ mg.} \end{aligned}$$

(b) Alternatively, at s.t.p. $\varrho_1 = 20.2/22.4$ kg m^{-3}, since 22.4 m^3 is the volume of one kilomole of neon, for which $M_r = 20.2$.

Then, using eqn. (15.3),

$$\varrho_2 = \varrho_1\left(\frac{T_1}{T_2}\right)\left(\frac{p_2}{p_1}\right) = \frac{20.2}{22.4} \times \frac{273}{293} \times \frac{800}{760} = 0.88 \text{ kg m}^{-3}.$$

Therefore in the volume of 10^{-5} m^3, the mass of neon $= 0.88 \times 10^{-5}$ kg $= 8.8$ mg.

15.2. Worked Example

A vessel contains a mass of 10 g of nitrogen at a pressure of 500 mmHg. Calculate the mass of oxygen which must be introduced at the same temperature to increase the pressure to 860 mmHg. Assume the relative molecular masses of nitrogen and oxygen to be 28 and 32 respectively.

Introduction

Consider two equal volumes of ideal gases A and B at the same temperature T and pressure p. Assuming Avogadro's law, these volumes contain equal numbers of molecules N even if the molecules have different masses.

If a certain number a of the A molecules are transferred to another equal and previously evacuated container, they will exert a pressure of $(a/N)p$ since pressure is proportional to mass (see eqn. (15.5)]. Alternatively, if a number b of the B molecules are transferred, they alone would exert a pressure $(b/N)p$. Now, if $a+b = N$, and both a and b are together in the container, with no chemical reaction, the total pressure of the mixture must be p according to Avogadro's law. But if $a+b = N$, $(a/N)p+(b/N)p=p$. Therefore the total pressure of a mixture of gases at a particular temperature is the sum of the partial pressures which the separate components of the mixture would exert if each occupied the volume alone at that temperature. This is the law of partial pressures.

Of course, the gases A and B need not be chemically different, but if they are identical the same result follows simply from the proportionality of pressure and mass, in eqn. (15.5).

Solution

From eqn. (15.5), $V/(TR) = m/(pM_r)$. But V, T, and R are the same for both the oxygen and the nitrogen.

$$\therefore \quad [m/(pM_r)]_O = [m/(pM_r)]_N,$$

where O refers to oxygen and N to nitrogen and the equation applies whatever units are used so long as they are the same on both sides. Thus m may be in grams and p in mmHg. Also from the law of partial pressures, $500+p_O = 860$.

$$\therefore\ P_O = 360 \quad \text{and} \quad m_O/(360\times 32) = 10/(500\times 28).$$

$$\therefore\ m_O = 360\times 32\times 10/(500\times 28) = 8.2 \text{ g}.$$

15.3. Calculate the universal gas constant from the volume of one kilomole of gas. [8.3×10^3 J K^{-1} $kmole^{-1}$.]

15.4. Calculate the relative molecular mass of carbon dioxide from the value of the universal gas constant and the density of 1.98 kg m^{-3} for carbon dioxide at s.t.p. [44.]

15.5. Calculate the number of molecules in 5.0 l of gas at a temperature of 100°C and pressure of 10 mmHg. Ref.: Avogadro's number and ideal gas volume. [1.3×10^{21}.]

15.6. A gas cylinder of volume 50 l. contains hydrogen at a pressure of 20 atm. and at a constant temperature of 15°C. Calculate the pressure to which the gas in the cylinder would drop if a mass of 30 g of the gas were released. Ref.: hydrogen. [13 atm.]

15.7. A one-litre flask contains air at s.t.p. How many molecules of nitrogen (relative molecular mass = 28) must be added to increase the mass by 1.0 g? What would be the final pressure if the final temperature were 27°C? Ref.: Avogadro's number and ideal gas volume. [2.1×10^{22}; 1500 mmHg.]

Saturated Vapour

15.8. Worked Example

A constant mass of air, contained over water, has a volume of 0.40 l. at a pressure of 270 mmHg and a temperature of 20°C. Calculate the volume when the temperature is raised to 40°C at constant pressure. Ref.: vapour pressure.

Introduction

At the free surface of a liquid some molecules have velocities great enough in suitable directions to enable them to escape from the surface against the forces of attraction to the remainder (see also ex. 8.25). If the space above the surface is closed, as in a stoppered bottle, not only are some molecules leaving the surface at any moment but also others are returning after collisions in this space. If there is sufficient liquid, the space in the bottle is "saturated" with vapour when dynamic equilibrium is reached, i.e. when vaporization and condensation occur at equal rates, so maintaining constant vapour density and pressure.

At higher temperatures, the average velocity increases, and more molecules at the surface have sufficient energy to leave. Then the rate of evaporation is increased and more vapour is formed before dynamic equilibrium is reached. Thus the "saturated vapour pressure" increases rapidly with temperature due not only to the higher temperature of the vapour but to its increased density.

In this way a saturated vapour differs from an ideal gas in which the number of molecules cannot be changed solely by changing the temperature. A difference is seen also at constant temperature if the volume of vapour is changed. Thus if the stopper were pressed in so as to reduce the volume of vapour, the density would increase only momentarily. The rate of condensation would increase until dynamic equilibrium had been restored with the original vapour density and pressure. Thus the saturated vapour pressure is the maximum that can be maintained above the liquid at a particular temperature, whatever the volume.

In a mixture of gas and saturated vapour, the total pressure

$$p = p_g + p_{sv}, \qquad (15.6)$$

where the gas pressure p_g may usually be considered to follow the ideal gas laws, and the pressure of saturated vapour p_{sv} does not. Equation (15.6) follows the law of partial pressures in that the component pressures are independent of each other, each being the same as if the other were absent.

Solution

It is assumed that all components of the air other than the water vapour are ideal gases. So, when the total pressure $p = 270$ mmHg and $p_{sv} = 17.5$ mmHg at 20°C, from the reference section,

$$p_g = 270-17.5 = 252.5 \text{ mmHg at } 20°\text{C}.$$

Similarly at 40°C, $p_g = 270-55.3 = 214.7$ mmHg.

But for the ideal gases, $p_{g1}V_1/T_1 = p_{g2}V_2/T_2$ from eqn. (9.10). In this case, $p_{g1} = 252.5$, $V_1 = 0.40$, $T_1 = 273+20 = 293$, $p_{g2} = 214.7$, V_2 is to be found, $T_2 = 273+40 = 313$.

$$V_2 = (p_{g1}V_1/T_1)(T_2/p_{g2})$$
$$= (252.5\times 0.40/293)(313/214.7) = 0.50 \text{ l}.$$

15.9. Worked Example

A cylinder of capacity 1.5 l. contains air saturated with water vapour at a temperature of 40°C and total pressure of 100 mmHg. Calculate the mass of water condensed at constant temperature (a) when the volume is halved, and (b) when the total pressure is doubled. Ref.: vapour density and pressure.

Introduction

The density of saturated vapour is determined by its temperature; the density cannot increase without becoming supersaturated or decrease without becoming unsaturated at that temperature. Supersaturation is possible under some circumstances but is not stable. Unsaturated vapours behave approximately as ideal gases up to the limit of saturation.

The density of a vapour can be calculated from Avogadro's law. Thus eqn. (15.4) gives the saturated vapour density

$$\varrho_{sv} = M_r p_{sv}/(TR), \qquad (15.7)$$

where $M_r = 18$ for H_2O since the relative atomic masses are $H = 1$ and $O = 16$.

It is useful to compare the densities of pure water vapour and dry air at the same p and T. Thus from eqn. (15.7), $\varrho_v = \varrho_a(M_{rv}/M_{ra})$, where v stands for pure vapour and a for dry air. M_{ra} has a mean value of 29 for dry air having the usual proportions of oxygen and nitrogen. Therefore up to saturation,

$$\varrho_v = \varrho_a(18/29) = 0.62\varrho_a. \qquad (15.8)$$

Note that the density of water vapour is not given directly by $\varrho_v = M_{rv}/V$, where V is the ideal gas volume at s.t.p., since pure water vapour cannot exist at s.t.p.; at the standard temperature of 0°C the saturation pressure is 4.6 mmHg, and at the standard pressure of 760 mmHg the temperature is 100°C.

Solution

(a) The reference table gives $P_{sv} = 55.3$ mmHg at 40°C, and $\varrho_a = 1.29$ kg m^{-3} at s.t.p. Therefore using eqns. (9.11) and (15.8),

$$\begin{aligned}\varrho_{sv} &= 0.62\times1.29\times(273/313)\times(55.3/760)\\ &= 0.051 \text{ kg m}^{-3} \text{ at } 55.3 \text{ mmHg and } 40°\text{C}.\end{aligned}$$

Since ϱ_{sv} is constant while the volume is reduced by 0.75×10^{-3} m^3, the mass condensed

$$= 0.051\times0.75\times10^{-3} = 38\times10^{-6} \text{ kg} = 38 \text{ mg}. \qquad \text{(a)}$$

(b) Assuming the "dry" components of the air to follow Boyle's law, the volume is inversely proportional to the pressure p_a, excluding the vapour pressure. Thus when the total pressure changes from 100 mmHg to 200 mmHg, p_a changes from (100−55.3) to (200−55.3) mmHg, i.e. from 44.7 to 144.7 mmHg, and the volume changes from 1.5×10^{-3} to $1.5\times10^{-3}\times44.7/144.7 = 0.46\times10^{-3}$ m^3. Therefore the volume is reduced by 1.04×10^{-3} m^3 and the mass condensed

$$= 0.051\times1.04\times10^{-3} = 53\times10^{-6} \text{ kg} = 53 \text{ mg}. \qquad \text{(b)}$$

15.10. Some air, saturated with water vapour, has a volume of 60 cm^3 at a pressure of 770 mmHg, and 90 cm^3 at 520 mmHg, with no change of temperature. Calculate the saturated vapour pressure at this temperature. [20 mmHg.]

15.11. A closed vessel contains air saturated with water vapour at a temperature of 75°C and a total pressure of 760 mmHg. Calculate the total pressure when the temperature is lowered to 30°C assuming the volume of air to remain constant. Ref.: vapour pressure. [440 mmHg.]

15.12. A mass of air, contained over water in a vessel at a constant temperature of 75°C, has a volume of 3.0 l. at a pressure of 760 mmHg. Calculate (a) the volume of air when the water level is lowered to decrease the pressure to 610 mmHg, and (b) the pressure when the volume is doubled. Ref.: vapour pressure. [(a) 4.4 l.; (b) 525 mmHg.]

15.13. The bulb of a constant volume gas thermometer contains air and water at 40°C. The pressure rises from 800 mmHg at 40°C to 1195 mmHg at 80°C. Calculate the saturated vapour pressure at 80°C, assuming the volume of liquid water to be negligible but sufficient to maintain saturation. Ref.: vapour pressure. [355 mmHg.]

15.14. The temperature of the water in question 15.12 is raised to 100°C, and the volume of air is increased to 5.0 l. Calculate (a) the new pressure, and (b) the percentage increase in the mass of water vapour. [(a) 1060 mmHg; (b) 310%.]

15.15. A bottle containing liquid carbon tetrachloride (CCl_4) encloses a volume of air of 1.0 l. above the liquid. Calculate the mass of carbon tetrachloride which evaporates when the temperature rises from 12°C to 23°C, assuming the saturated vapour pressure to rise from 60 mmHg to 100 mmHg, and the relative atomic masses C = 12 and Cl = 35.5. Ref.: ideal gas and mercury. [3.2 g.]

Relative Humidity

15.16. Worked Example

Some air at a temperature of 60°C and a pressure of 740 mmHg has a density of 0.99 kg m^{-3}. Calculate (a) the relative humidity, and (b) the dew point. Ref.: densities and vapour pressures.

Introduction

The humidity of the atmosphere, or of any gas, may be stated by the density of the water vapour which it contains. However, the net rate of evaporation of water into a humid atmosphere depends not only upon the absolute density of the water vapour already in the atmosphere but also upon the degree to which this approaches saturation at the particular temperature. Thus if the temperature drops with no change of absolute humidity the atmosphere seems to become more humid because of the increase of relative humidity, which is defined as the ratio ϱ_v/ϱ_{sv}, where ϱ_v is the actual density of the vapour at the particular temperature, and ϱ_{sv} is the saturation density at the same temperature.

Assuming that Avogadro's law applies, the vapour density is proportional to the vapour pressure p_v at all values up to saturation, as in eqn. (15.7). Then

$$\text{relative humidity} = \varrho_v/\varrho_{sv} = p_v/p_{sv}. \qquad (15.9)$$

If the temperature of the atmosphere falls sufficiently, water vapour may be deposited as dew or frost. The dew point is the temperature to which the particular atmosphere must be cooled at constant pressure to become saturated and so to be on the point of depositing dew. Thus the pressure of a vapour always equals the saturation pressure at its dew point.

$$\therefore \text{ relative humidity} = \frac{p_v}{p_{sv}} = \frac{p_{sv} \text{ at dew point}}{p_{sv} \text{ at actual temperature.}} \qquad (15.10)$$

Solution

(a) When total $p = 740$ mmHg, $p_a = 740 - p_v$. But from the reference section, $\varrho_a = 1.29$ kg m^{-3} at s.t.p.

$$\therefore \text{ at } 60°\text{C}, \quad \varrho_a = 1.29 \times (273/333) \times (740 - p_v)/760$$
$$= 1.39 \times 10^{-3}(740 - p_v).$$

Also, from eqn. (15.8), $\varrho_v = 0.62 \times \varrho_a$ for the same temperature and pressure, up to saturation.

$$\therefore \text{ at } 60°\text{C}, \quad \varrho_v = 0.62 \times 1.29 \times (273/333) \times p_v/760$$
$$= 0.864 \times 10^{-3} p_v.$$

But $$\varrho_a + \varrho_v = 0.99.$$

$$\therefore \; 1.39 \times 10^{-3}(740 - p_v) + 0.864 \times 10^{-3} p_v = 0.99.$$

$$\therefore \; p_v = \frac{(1.39 \times 10^{-3} \times 740) - 0.99}{(1.39 - 0.864) \times 10^{-3}} = 76 \text{ mmHg}.$$

Then, since p_{sv} at 60°C is given in the reference section as 149 mmHg, eqn. (15.10) gives

$$\text{relative humidity} = 76/149 = 51\%. \qquad \text{(a)}$$

(b) If the reference values of saturation pressure are plotted as a graph against temperature, it is found that the temperature for a pressure of 76 mmHg is approximately 46°C.

This is the dew point. (b)

15.17. Calculate the density of air at a temperature of 30°C and pressure of 760 mmHg when the relative humidity is 60%. Ref.: air and vapour densities and vapour pressure. [1.15 kg m^{-3}.]

15.18. Some air is saturated with water vapour at 60°C. Calculate the relative humidity at 90°C when (a) the pressure, and (b) the volume is maintained constant with no increase of water content. Ref.: vapour pressures. [(a) 28%; (b) 31%.]

15.19. Some humid air at 60°C and 300 mmHg has a density of 0.368 kg m^{-3}. Calculate the relative humidity. Ref.: air and vapour densities and vapour pressure. [0.6.]

15.20. A certain mass of moist air at a temperature of 60°C and a pressure of 760 mmHg contains 3% of moisture by weight. Calculate the relative humidity of the air. Ref.: air and vapour densities and vapour pressure. [24%.]

Pressure and Kinetic Energy

15.21. Worked Example

Calculate the thermal kinetic energy of translation and the r.m.s. molecular velocity of a mass of 100 g of nitrogen at a temperature of 50°C, assuming the density of nitrogen at s.t.p. to be 1.25 kg m^{-3}.

Introduction

The following theory is based on several simplifying assumptions: (i) the gas consists of a very large number of very small identical molecules moving continually in a volume of space much greater than their own total volume; (ii) the only forces acting on the molecules are very short range repulsions during brief periods of collision, attraction and weight being negligible; (iii) the kinetic energies of the molecules are due to linear movement only, and may be redistributed locally by collisions, although the overall distribution and the total energy remains constant; (iv) the walls of the container present perfectly smooth and impenetrable barriers from which the molecules of gas rebound with no change of speed.

Now consider a rectangular box of dimensions x, y, z, containing gas molecules, each of mass m_1. If one of these strikes one wall, as in Fig. 15.1, the perpendicular velocity component u_x is reversed in the reaction to the impulse on the wall, while the speed u is

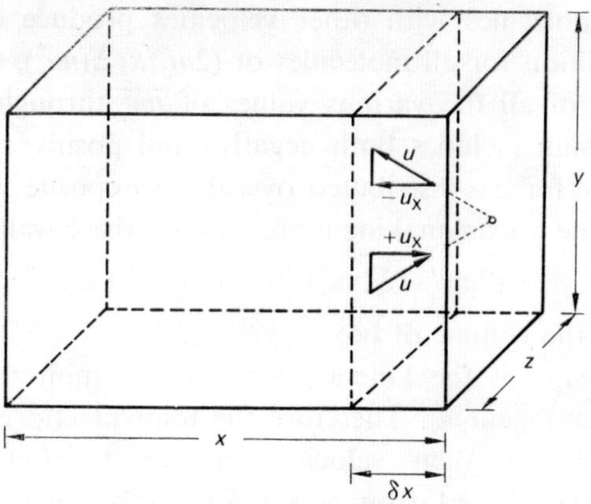

FIG. 15.1. Ex. 15.21.

unchanged. But the impulse equals the change of momentum (see ex. 3.11). Thus when the momentum changes from $+m_1u_x$ through zero to $-m_1u_x$, the total change is $-2m_1u_x$ and the impulse on the wall is

$$+2m_1u_x. \qquad (15.11)$$

Many similar impulses at different parts of the wall are due to many other molecules arriving with velocity u_x also. All such molecules in the volume $yz\,\delta x$ (see Fig. 15.1) may be considered to reach the wall without striking other molecules if δx is very small. All the resulting impulses occur within a time $\delta t = \delta x/u_x$. But if n molecules with velocity u_x are distributed throughout the whole volume (xyz), the number of these in the volume $(yz\,\delta x)$ is $n\delta x/x$. Therefore the mean frequency of impulses, each given by (15.11),

$$= (n\,\delta x/x)/\delta t = (n\,\delta x/x)/(\delta x/u_x) = nu_x/x. \qquad (15.12)$$

Now the mean force on the wall is the total impulse per unit time. Therefore, from eqns. (15.11) and (15.12) the mean force due to these impulses

$$= (2m_1u_x)(nu_x/x) = (2m_1/x)(nu_x^2).$$

Other molecules with other velocities produce other forces, with a resultant for all molecules of $(2m_1/x)\,\Sigma(nu_x^2)$, where $\Sigma(nu_x^2)$ is the sum of all the various values of nu_x^2 throughout the gas. Since this sum includes both negative and positive values of u_x, the resultant force is distributed over both opposite walls perpendicular to the x-axis, making a pressure on these walls

$$p = (2m_1/x)\,\Sigma(nu_x^2)/2yz = (m_1/V)\,\Sigma(nu_x^2), \qquad (15.13)$$

where V is the volume of box.

Now $\frac{1}{2}m_1u_x^2$ is the kinetic energy of one molecule due to its velocity component u_x. Therefore the total kinetic energy of all molecules due to their velocity components along the x-axis equals $\frac{1}{2}m_1\Sigma(nu_x^2)$ and, from eqn. (15.13), this equals

$$pV/2. \qquad (15.14)$$

Since there is no draught through the box, no particular direction of movement is favoured, and the total kinetic energy of translation (linear motion) is due equally to components of velocity along all three axes x, y, z. Therefore total kinetic energy of translation in all directions,

$$E_k = 3pV/2. \qquad (15.15)$$

Thus pV is a constant product for this ideal gas at constant temperature, assuming that E_k is constant at constant temperature (see Boyle's law).

If N is the total number of molecules, the mean energy per molecule

$$\bar{E}_k = (n_1\,\tfrac{1}{2}m_1u_1^2 + n_2\,\tfrac{1}{2}m_1u_2^2 + \ldots)/N.$$

But $\sqrt{(n_1u_1^2+n_2u_2^2+\ldots)/N}$ is the "root mean square" velocity u_{rms}, such that

$$\bar{E}_k = \tfrac{1}{2}m_1u_{\text{rms}}^2. \qquad (15.16)$$

Therefore eqn. (15.15) gives

$$pV = \tfrac{2}{3}E_k = \tfrac{2}{3}N\bar{E}_k$$
$$= \tfrac{1}{3}Nm_1u_{\text{rms}}^2. \qquad (15.17)$$

Finally, Nm_1/V equals the density ϱ. Therefore eqn. (15.17) gives

$$p = \tfrac{1}{3}\varrho u_{\text{rms}}^2. \qquad (15.18)$$

Solution

(a) In the particular question, the density at 50°C and 760 mmHg is

$$\varrho = 1.25\times 273/323 = 1.056 \text{ kg m}^{-3}$$

and

$$p = 1.013\times 10^5 \text{ N m}^{-2}.$$

Therefore for a mass of 0.10 kg,

$$V = 0.10/1.056 = 0.0946 \text{ m}^3.$$

Then eqn. (15.15) gives

$$E_k = \tfrac{3}{2}pV = \tfrac{3}{2}(1.013\times 10^5\times 0.0946) = 1.44\times 10^4 \text{ J.} \qquad \text{(a)}$$

(b) From eqn. (15.18),

$$u_{\text{rms}} = \sqrt{(p/\varrho)}\sqrt{(3\times 1.013\times 10^5/1.056)} = 540 \text{ m s}^{-1}. \qquad \text{(b)}$$

15.22. Calculate the mean velocity and the r.m.s. velocity of hydrogen molecules at a temperature of 1000°C from the following approximate data. The molecules are divided into five groups of equal numbers. In each group all velocities are considered to be equal. The five velocities are taken to be the following multiples of 10^3 m s^{-1}: 1.7, 2.7, 3.5, 4.4, 5.9. [3.6×10^3 and 3.9×10^3 m s^{-1}.]

15.23. Hydrogen molecules at a temperature of 1000°C may be divided approximately into seven groups in each of which all velocities are considered to be equal. The seven values of velocity, one for each group, are given by $n\times 10^3$ m s^{-1}, where n = all integers from 1 to 7 in succession. The numbers of molecules in the groups are in the ratios 4 : 11 : 15 : 14 : 10 : 5 : 2 respectively. Calculate both the mean velocity and the r.m.s. velocity. [See ex. 15.22.]

15.24. A spherical vessel of radius 10 cm contains 10^{23} molecules of CO_2 having an r.m.s. velocity of 770 m s^{-1}. Calculate the pressure exerted by the molecules, assuming that the mass of the hydrogen atom = 1.67×10^{-27} kg and that the relative atomic masses of C and O are 12 and 16 respectively. [3.5×10^5 N m^{-2}.]

15.25. Calculate the r.m.s. velocity of neon molecules at 0°C given the density of neon to be 0.90 kg m^{-3} at s.t.p. [580 m s^{-1}.]

Temperature and Kinetic Energy

15.26. Worked Example

Calculate Boltzmann's constant from the data of the reference section, and hence calculate the mean thermal kinetic energy of translation of ideal gas molecules at a temperature of 500°C.

Introduction

From eqn. (15.15) it follows that the mean kinetic energy of translation of ideal gas molecules $\bar{E}_k = (\frac{3}{2})(pV/N)$, where N is the number of molecules in volume V at pressure p. But for an ideal gas, $pV = (Rm/M_r)T$, where R is the universal gas constant, m is the mass of gas, M_r is the relative molecular mass, and T is the Kelvin temperature.

$$\therefore \ \bar{E}_k = (\tfrac{3}{2})(Rm/NM_r)T. \qquad (15.19)$$

Also, from eqn. (15.1), $m = NM_r m_c/12,$ (15.20)

where m_c is the mass of one molecule of carbon-12.

$$\therefore \ \bar{E}_k = (\tfrac{3}{2})(Rm_c/12)T, \qquad (15.21)$$

where R and m_c are constants.

Thus the absolute temperature on the ideal gas scale is proportional to the mean kinetic energy of translation of ideal gas molecules of any mass.

Note the close connection between the p and T. Thus for any ideal gas the pressure equals two-thirds of the total E_k of translation per unit volume, whereas the temperature is a measure of the mean of this energy per molecule. The constant volume gas thermometer uses the proportionality of p and T as a measure of T.

For 1 kilomole of gas, $m = M_r$, and eqn. (15.20) gives the number of molecules per kilomole, called Avogadro's number, $N_A = 12/m_c$. Then eqn. (15.21) may be written

$$\bar{E}_k = (\tfrac{3}{2})(R/N_A)T = (\tfrac{3}{2})kT, \qquad (15.22)$$

where $R/N_A = k$, called Boltzmann's constant.

Solution

From the reference data,

$$k = 8.31\times10^3/(6.02\times10^{26})$$

$$= 1.38\times10^{-23}\text{ J K}^{-1}.$$

Then for $T = 500+273$ K, eqn. (15.22) gives

$$\bar{E}_k = (\tfrac{3}{2})(1.38\times10^{-23})773 = 1.6\times10^{-20}\text{ J}.$$

15.27. Calculate the mean and the total translational kinetic energy of molecules in 2.0 g of helium at 0°C, assuming the relative molecular mass of helium to be 4, and using the values of the universal gas constant and Avogadro's number. [5.7×10^{-21} J; $.7\times10^3$ J.]

15.28. Calculate the total thermal kinetic energy of translation of one kilomole of ideal gas at s.t.p. [3.4×10^6 J.]

15.29. Given the r.m.s. velocity of hydrogen molecules at 0°C to be 1.85×10^3 m s^{-1}, calculate the value for oxygen molecules at 00°C. Assume the relative molecular masses of H_2 and O_2 to be and 32 respectively. [540 m s^{-1}.]

CHAPTER 16

FLUID FLOW

Flow and Work

16.1. Worked Example

A cylindrical pipe of diameter 10 cm leads into two cylindrical pipes each of diameter 2.5 cm. Liquid which fills the pipes, flows steadily at a mean velocity of 4.0 mm s^{-1} in the main pipe and then divides equally into the smaller pipes. Calculate (a) the rate of flow, and (b) the mean velocity in each of the smaller pipes.

Introduction

Since the quantity of liquid cannot increase or decrease at the junction when the system of pipes is full, the mass arriving at the junction must equal the mass leaving in the same time. Also, if the change of density with pressure is negligible, as may be assumed for a liquid but not necessarily for a gas, the volumes arriving and leaving must be the same in the same time. Then the rate of flow in units of volume or mass per unit time is twice as great in the main pipe as in each of the others.

The velocity of the liquid, as distinct from its rate of flow, is measured in units of distance per unit time. This depends upon the cross-section area of the pipe as well as upon the rate of flow. Thus if the pipe, as in Fig. 16.1, has cross-section area A, and the volume V of liquid occupying the length l from X to Y moves into the position Y to Z at a velocity v, it takes a time l/v, and therefore

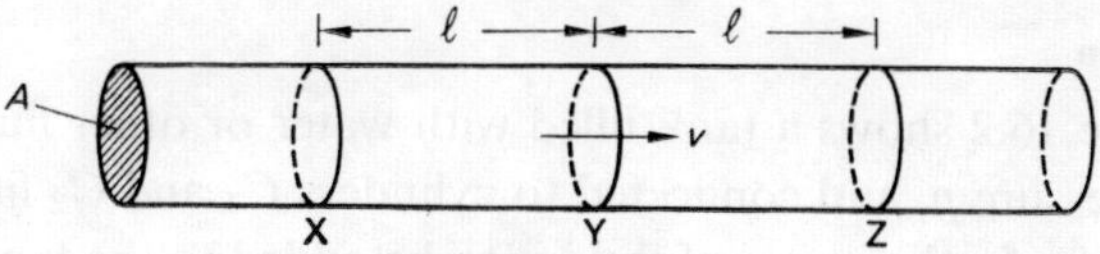

FIG. 16.1. Fluid flow in a pipe (ex. 16.1).

flows at the rate $V/(l/v)$. But $V = Al$ for a pipe of constant bore. Therefore the rate of flow in terms of volume per unit time

$$= Al/(l/v) = Av, \tag{16.1}$$

where v is the velocity in terms of distance per unit time.

Since a liquid tends to drag at the wall of a pipe, its velocity is not uniform over the cross section but has the mean value v used above.

Solution

(a) In the pipe of diameter 0.10 m, the rate of flow given by eqn. (16.1), using the metre as the unit of length, is

$$Av = (\pi \times 0.10^2/4) \times (4.0 \times 10^{-3}) = \pi \times 10^{-5}\ \text{m}^3\ \text{s}^{-1}.$$

Then in each smaller pipe the rate is half of this, i.e.

$$\tfrac{1}{2}\pi \times 10^{-5} = 1.6 \times 10^{-5}\ \text{m}^3\ \text{s}^{-1}. \tag{a}$$

(b) The velocity in each smaller pipe is given by

$$\begin{aligned} &(\text{rate of flow})/(\text{cross-section area}) \\ &= (\tfrac{1}{2}\pi \times 10^{-5})/(\tfrac{1}{4}\pi \times 2.5^2 \times 10^{-4}) \\ &= 3.2 \times 10^{-2}\ \text{m s}^{-1} \end{aligned} \tag{b}$$

16.2. WORKED EXAMPLE

A hydraulic system includes a closed tank of water at the base of which a constant gauge pressure of 2000 mmHg is maintained. Calculate (a) the work required to feed a volume of 1.5×10^4 l. of water slowly to the tank at its base, and (b) the power transmitted by a flow at a rate of 0.30 l. s^{-1} at this pressure.

Introduction

Figure 16.2 shows a tank filled with water or other fluid under gauge pressure p, and connected to cylinders C_1 and C_2 fitted with pistons. If a further mass of fluid is to be added to the tank from a supply at atmospheric pressure, it may be run into the cylinder C_1 through valve X_2 when valve X_1 is closed, and then forced through X_1 by applying a suitable force F_1 to the piston of C_1 when X_2 is closed and X_1 open.

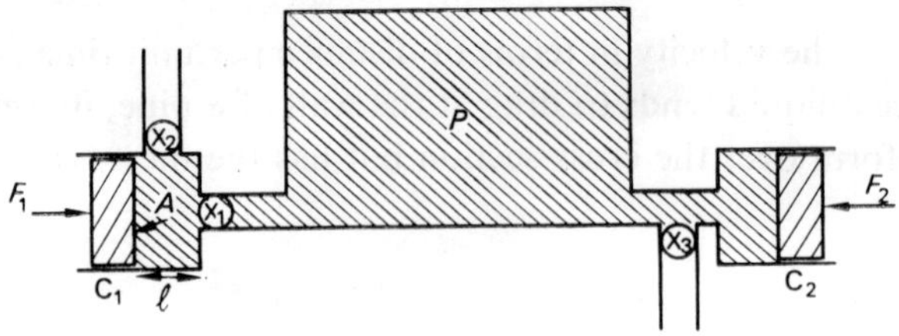

FIG. 16.2. Ex. 16.2.

If acceleration and friction are negligible F_1 must equal the force pA of the fluid acting on the piston area A. When the piston is driven through a distance l, the work done by the piston on the fluid equals

$$F_1 l = pAl = pV, \tag{16.2}$$

where V = volume of fluid forced through X_1, and p = the gauge pressure in the tank. Note that p equals the difference between the absolute pressure in the tank and the atmospheric pressure of the supply, just as each force F_1 and F_2 must equal the difference between the forces due to these same pressures acting on the inner and outer faces of the particular piston.

The product pV is sometimes called the pressure energy, although this is not a form of energy, stored in a volume V of fluid. It is energy transmitted by the flow of fluid, as seen by considering the cylinder C_2 closed by a piston on which force F_2 acts to maintain the constant pressure p. Then, if X_3 is closed, the work done by F_1 equals the work done against F_2, i.e. energy pV is transmitted by the fluid from the input at one piston to the output at the other, and does not remain in the fluid. Similarly, if X_3 is opened, energy may be transmitted by the flow under pressure through X_3. Then

$$\text{output power} = (\text{energy transmitted})/(\text{time taken})$$
$$= pV/t = (\text{pressure})\times(\text{rate of flow}). \quad (16.3)$$

Solution

(a) In M.K.S. units,

$$p = h\varrho g = 2000\times10^{-3}\times1.36\times10^{4}\times9.81$$
$$= 2.67\times10^{5}\ \text{N m}^{-2}.$$

Also $$V = 1.5\times10^{4}\times10^{-3} = 15\ \text{m}^{3}.$$

Then from eqn. (16.2),

$$\text{work required} = pV = 2.67\times10^{5}\times15 = 4.0\times10^{6}\ \text{J}. \quad \text{(a)}$$

(b) From eqn. (16.3),

$$\text{power transmitted} = p(V/t)$$
$$= 2.67\times10^{5}\times0.30\times10^{-3} = 80\ \text{W}. \quad \text{(b)}$$

16.3. Water flowing at a constant rate from a tap falls freely in a continuous stream which at one point has a cross-section diameter of 3.0 mm. If the water fills a bottle of capacity 0.50 l. in a time of 2.5 min, calculate the velocity of the water at the section referred to. [47 cm s^{-1}.]

16.4. Three pipes of internal diameters 2.0 cm, 4.0 cm, and 6.0 cm respectively, are connected at a junction. Oil of density 920 kg m^{-3} flows at a velocity of 0.50 m s^{-1} toward the junction in each of the smaller pipes. Calculate (a) the mean velocity of the oil in the largest pipe, and (b) the rate of flow in kg h^{-1} in each of the smaller pipes. [(a) 0.28 m s^{-1}; (b) 5.2×10^{2} kg h^{-1}, 2.1×10^{3} kg h^{-1}.]

16.5. The figures given overleaf give the velocity of liquid in a pipe of diameter 5.0 cm at the stated intervals of time. Estimate, with the aid of a graph, the mean rate of flow of the liquid, assuming a smooth variation of velocity between the given values.

Velocity ($cm\ s^{-1}$)	33 42 54 58 51 39 30
Time (min)	0 5 10 15 20 25 30

[900 $cm^3\ s^{-1}$.]

16.6. Calculate the gauge pressure required in a hose pipe to force a jet of water to a height of 10 m, assuming negligible resistance to flow through the nozzle. Ref.: density. [9.8×10^4 N m^{-2}.]

16.7. Calculate the rate of flow in litres per hour given by a fire pump of 60% efficiency drawing a power of 15 kW and pumping water to a height of 50 m. Calculate also the pressure in the water at the nozzle of the hose. [6.6×10^4 l. h^{-1}; 4.9×10^5 N m^{-2}.]

Bernoulli's Equation

16.8. Worked Example

The diameter of a Venturi meter varies from 10 cm at the entrance to 4.0 cm at the throat. Assuming ideal flow, (a) calculate the rate of flow of water through the meter when the gauge pressures at the entrance and throat are 25.0 cmH_2O and 20.0 cmH_2O respectively. (b) Calculate also the throat diameter of a Venturi tube with entrance diameter of 10 cm which reduces the pressure at the throat to atmospheric when the rate of flow and the main pressures are as in (a).

Introduction

In a Venturi tube (Fig. 16.3), the change of cross-section is designed to change the velocity of the fluid passing through and so to change the pressure. When the reduced pressure obtained at a side entrance to the throat is below atmospheric, it may be used as in a carburettor or suction pump. Also measurement of the decrease of pressure, as in a Venturi meter, enables the rate of flow

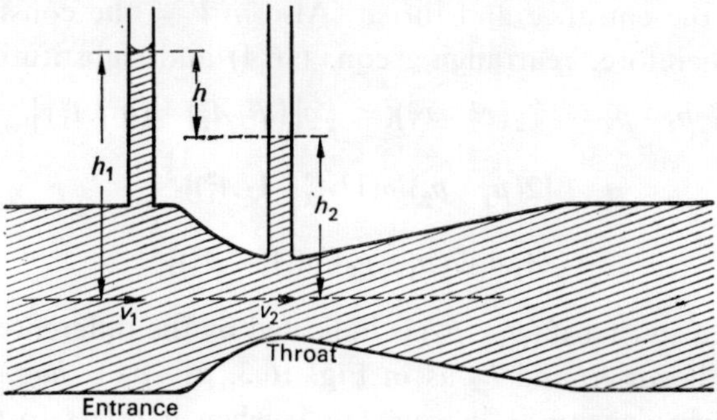

FIG. 16.3. The Venturi tube (ex. 16.8).

to be calculated by using Bernoulli's equation (16.8) or a reduced form of the equation obtained as follows.

Consider a small volume V of water of mass m entering the tube with velocity v_1 and an equal quantity passing through the throat with velocity v_2, assuming constant streamline flow, i.e. flow having no turbulence or irregular swirling. The input energy due to this flow through the entrance is in two parts: (1) kinetic energy $\frac{1}{2}mv_1^2$, and (2) energy p_1V transmitted by flow of volume V at the entrance pressure p_1, in accordance with eqn. (16.2).

$$\therefore \text{ energy input } = \tfrac{1}{2}mv_1^2 + p_1V.$$

Similarly, the energy passed through the throat in the same time is $\frac{1}{2}mv_2^2 + p_2V$, where p_2 is the pressure in the throat, and the density is assumed to be constant. Then, if there are no other changes of energy such as heat or potential energy,

$$\tfrac{1}{2}mv_1^2 + p_1V = \tfrac{1}{2}mv_2^2 + p_2V. \qquad (16.4)$$

Alternatively, it may be considered that the gain of kinetic energy $\frac{1}{2}mv_2^2 - \frac{1}{2}mv_1^2$ equals the work $(p_1 - p_2)V$ done by the pressure difference $(p_1 - p_2)$ as volume V passes through.

Now, from eqn. (16.1), the rate of flow in terms of volume per unit time $q = A_1v_1 = A_2v_2$, where A_1 and A_2 are the cross-section areas of the entrance and throat. Also m/V = the constant density ϱ. Therefore, rearranging eqn. (16.4) and substituting gives

$$p_1-p_2 = \tfrac{1}{2}\varrho(v_2^2-v_1^2) = \tfrac{1}{2}\varrho[(q^2/A_2^2)-(q^2/A_1^2)]. \qquad (16.5)$$

Then

$$q = [2(p_1-p_2)/\varrho(1/A_2^2-1/A_1^2)]^{\frac{1}{2}}$$

$$= A_1A_2[2(p_1-p_2)/\varrho(A_1^2-A_2^2)]^{\frac{1}{2}}. \qquad (16.6)$$

Note also that if (p_1-p_2) is indicated by the difference h of the pressure heads h_1 and h_2 as in Fig. 16.3, $p_1-p_2 = h\varrho g$, where ϱ is the same density as in eqn. (16.5) when it refers to the same liquid.

In practice the rate of flow is found to be somewhat less than that given by eqn. (16.6) if the cross-section area changes too rapidly, but here the difference is considered to be negligible in a well-designed tube. In the extreme case of an abrupt change of diameter, made by fitting a transverse "orifice plate" in a pipe, as in Fig. 16.4, the rate of flow may be only about 65% of that of

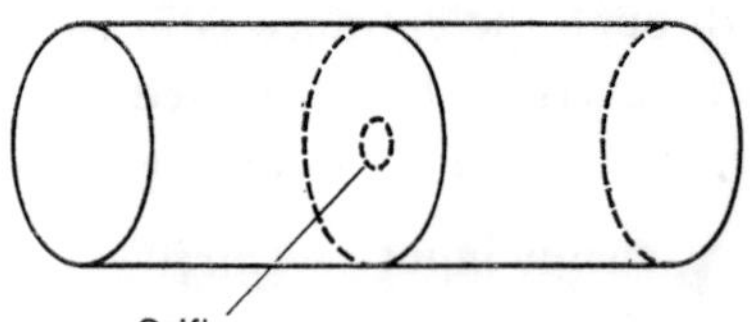

FIG. 16.4. Pipe with an orifice plate (ex. 16.8).

the corresponding ideal Venturi tube having the same maximum and minimum diameters. The orifice is then said to have a discharge coefficient of about 65%.

Solution

(a) For the given diameters,

$$A_1 = \pi 0.10^2/4 = 7.86\times10^{-3}\ \text{m}^2,$$
$$A_2 = \pi 0.04^2/4 = 1.26\times10^{-3}\ \text{m}^2.$$

Since the pressures are given in terms of the same liquid as in the tube, the density may be eliminated by substituting $p_1 - p_2 = (h_1 - h_2)\varrho g$ in eqn. (16.6):

$$q = A_1 A_2[2g(h_1 - h_2)/(A_1^2 - A_2^2)]^{\frac{1}{2}}$$
$$= 7.86\times1.26\times10^{-6}[2\times9.81\times5.0\times10^{-2}/(7.86^2-1.26^2)\times10^{-6}]^{\frac{1}{2}}$$
$$= 9.9\times10^{-6}[0.981\times10^6/(61.8-1.6)]^{\frac{1}{2}}$$
$$= 9.9\times10^{-6}\times1.28\times10^2 = 1.26\times10^{-3}\ \text{m}^3\ \text{s}^{-1}$$
$$= 1.3\ \text{l. s}^{-1}. \qquad \text{(a)}$$

(b) For the gauge pressure p_2 to be zero and other factors except A_2 unchanged, eqn. (16.5) gives

$$p_1 = \tfrac{1}{2}\varrho[(q^2/A_2^2) - (q^2/A_1^2)].$$

Therefore

$$25\times10^{-2}\times10^3\times9.81 = \tfrac{1}{2}\times10^3\times1.26^2\times10^{-6}[(1/A_2^2)-(10^6/61.8)]^{\frac{1}{2}}$$
$$\therefore\ 1/A_2^2 = (5.0\times10^5\times9.81/1.26^2)+(10^6/61.8)$$
$$= 3.09\times10^6\times0.016\times10^6.$$

Now $A_2 = \pi d_2^2/4$, where d_2 is the required diameter,

$$\therefore\ d_2^4 = 16/(\pi^2\times3.11\times10^6) = 52\times10^{-8}.$$
$$\therefore\ d_2 = 2.7\times10^{-2}\ \text{m}. \qquad \text{(b)}$$

16.9. Worked Example

An open tank of water, filled to a depth of 1.5 m, is set with its base at a height of 3.0 m above floor level. Water runs out through a small hole in the base. Calculate the velocity of the water (a) as it leaves the hole, and (b) as it reaches the floor.

Introduction

To establish Bernoulli's equation, consider a tube filled with fluid, shown in part in Fig. 16.5. If a small mass m of fluid of volume V enters the tube along the stream line through A at height

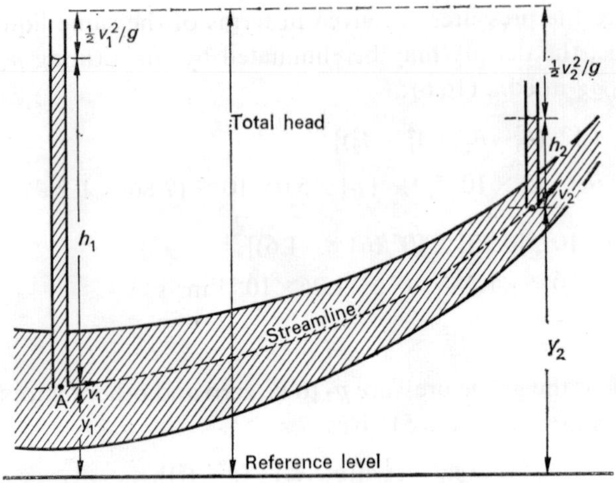

FIG. 16.5. To establish Bernoulli's equation.

y_1 with pressure p_1 and velocity v_1, an equal mass leaves at y_2 with pressure p_2 and velocity v_2 which are different due to the change of height and cross-section. Then, as in ex. 16.8, the total input of energy may be equated to the total output in the same time, assuming streamline flow and no gain or loss of energy such as heat. The resulting equation is like (16.4) but with a potential energy term on each side because of the different levels y_1 and y_2:

$$\tfrac{1}{2}mv_1^2+p_1V+mgy_1 = \tfrac{1}{2}mv_2^2+p_2V+mgy_2. \tag{16.7}$$

Then, since $m/V = \varrho$, which may be considered constant for a liquid,

$$\tfrac{1}{2}\varrho v_1^2+p_1+\varrho gy_1 = \tfrac{1}{2}\varrho v_2^2+p_2+\varrho gy_2. \tag{16.8}$$

This is a form of Bernoulli's equation with each term in dimensions of pressure, and is applicable to a steady streamline of fluid of constant density flowing in any direction, when the resistance is negligible. The terms $\frac{1}{2}\varrho v^2$ may be called dynamic pressures, since if the liquid were at rest, they would be zero, and the remaining terms would show the more elementary relationship of static pressure difference to difference of height, i.e. $p_1 - p_2 = (y_2 - y_1)\varrho g$.

An alternative form of Bernoulli's equation in terms of "head" may be obtained from eqn. (16.8) by substituting $p = h\varrho g$ and dividing through by ϱg:

$$v_1^2/2g + h_1 + y_1 = v_2^2/2g + h_2 + y_2. \qquad (16.9)$$

Here each term has the dimension of height, as shown in Fig. 16.5, and may be called a head: h is a pressure head equal to the height of liquid of the same ϱ in a barometer indicating the pressure p; y is a gravitational head measured from a fixed reference level; $v^2/2g$ is a velocity head equal to the distance required to reach the velocity v in falling freely from rest. Equation (16.9) shows the total head to be constant.

Solution

(a) It is assumed that the water leaving the hole, as in Fig. 16.6, has negligible turbulence, and has negligible velocity v_1 just inside

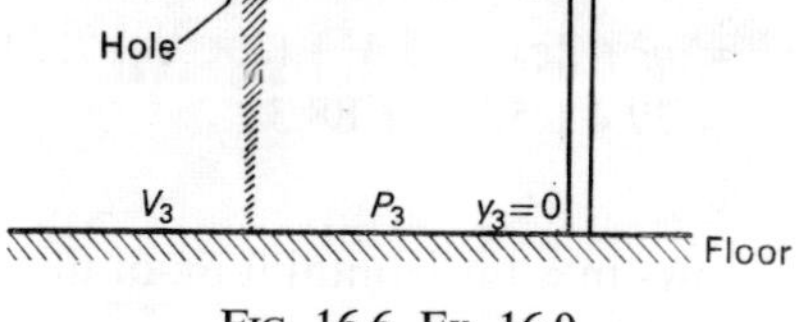

FIG. 16.6. Ex. 16.9.

the tank compared with v_2 just outside. Also y_1 and y_2 have negligible difference, and p_1 and p_2, or h_1 and h_2, differ due to the hydrostatic pressure at the base of the tank.

Then, substituting $v_1 = 0, y_1 - y_2 = 0, h_1 - h_2 = d$ in eqn. (16.9),

$$v_2^2 = 2gd = 2\times 9.81\times 1.5 = 29.4.$$

$$\therefore\ v_2 = 5.4 \text{ m s}^{-1}. \qquad \text{(a)}$$

Alternatively, it may be noted that, as the water leaves the hole, the level in the tank falls, so that the gain of kinetic energy equals the loss of potential energy of the water. The loss of potential energy is equivalent to the transfer of mass m from the level of the surface to the level of the hole since no increase or decrease of mass occurs at other levels. Thus the decrease of potential energy mgd equals the gain of kinetic energy $\frac{1}{2}mv_2^2$, again assuming negligible velocity inside the tank. Therefore $v_2^2 = 2gd$, as before.

(b) Using eqn. (16.9) again, and referring to Fig. 16.6, the equation relating levels 1 and 3 is

$$v_1^2/2g + h_1 + y_1 = v_3^2/2g + h_3 + y_3,$$

where $h_1 - h_3 = d$, as in (a), and $v_1 = 0$, $y_3 = 0$.

$$\therefore\ v_3^2 = 2g(d+y_1) = 2\times 9.81\times 4.5 = 88.3.$$

$$\therefore\ v_3 = 9.4 \text{ m s}^{-1}. \qquad \text{(b)}$$

Again, as for (a), an alternative method is to consider the loss of potential energy and gain of kinetic energy without considering the intermediate process of work done by the pressure.

A further alternative is to consider free fall for a distance $s = 3.0$ m under acceleration $g = 9.81$ m s^{-2}, commencing with velocity $u = 5.4$ m s^{-1} from (a). Then, using eqn. (1.3),

$$v^2 = u^2 + 2gs = 5.4^2 + (2\times 9.81\times 3.0)$$
$$= 29.4 + 58.9 = 88.3$$

(as before).

16.10. Water flows in a horizontal hose of diameter 2.0 cm at a pressure of 1.0×10^5 N m^{-2}, and issues from a nozzle as a jet of minimum diameter 0.50 cm. Calculate the mean velocity of water at this cross-section of the jet, assuming negligible resistance. Ref.: density. [14 m s^{-1}.]

16.11. Estimate the depth of water in a tank when the rate of flow through a hole of 1.0 mm diameter in the base is 2.0 $cm^3\ s^{-1}$. Assume a discharge coefficient of 0.60. Ref.: density. [0.92 m.]

16.12. A Pitot tube consists of a nozzle mounted to face an oncoming stream of fluid and connected to a manometer which indicates the difference of pressure between the stationary fluid in the nozzle and the fluid streaming past it. When the tube is set facing along the axis of a water pipe of diameter 25 cm, in which the mean velocity is 80% of the axial velocity, the manometer indicates a pressure difference of 10 cmH_2O. Calculate the rate of flow of the water in the pipe. [$5.5\times10^{-2}\ m^3\ s^{-1}$.]

16.13. A liquid of density 900 kg m^{-3} flows through a Venturi meter having inlet and throat diameters of 20 cm and 8.0 cm respectively. Assuming ideal flow, calculate the rate of flow of the liquid when the differential pressure produced is 10 cmH_2O. Ref.: water. [$7.5\times10^{-3}\ m^3\ s^{-1}$.]

16.14. An orifice plate mounted across a pipe of diameter 5.0 cm has a central orifice of diameter 2.0 cm. Assuming a discharge coefficient of 0.65, calculate the rate of flow of a fluid of density 900 kg m^{-3} when the pressure drop due to the orifice is 40 mmHg. Ref.: mercury. [$7.1\times10^{-4}\ m^3\ s^{-1}$.]

16.15. A vertical pipe of internal diameter 3.0 cm leads into a narrower section of 2.0 cm diameter. Calculate the gauge pressure in the pipe 10 cm above the narrow section when water flows at a rate of $3.0\times10^{-4}\ m^3\ s^{-1}$ with a gauge pressure of $5.0\times10^3\ N\ m^{-2}$ in the entrance to the narrow section. Neglect resistance to flow. Ref.: density. [$4.4\times10^3\ N\ m^{-2}$.]

Viscosity

16.16. Worked Example

When equal volumes of glycerine and castor oil are tested separately in a U-tube viscometer, the measured times of flow are 62 s and 97 s respectively. Assuming the densities of the liquids to be 1260 kg m^{-3} and 960 kg m^{-3} respectively, and the viscosity of the glycerine to be 0.83 kg m^{-1} s^{-1} at its temperature during the test, calculate (a) the dynamic viscosity, and (b) the kinematic viscosity of the castor oil.

Introduction

Viscosity is the tendency of a fluid to resist movement of one part of it relative to another part in contact with it. It is similar to sliding friction between solid surfaces in that forces of viscosity oppose the sliding of one layer of fluid relative to a parallel layer, but viscosity occurs throughout the fluid. An important surface effect is the frequent adhesion of a liquid to a solid, e.g. rain adheres to a window pane after most has run off.

Thus, when a stream of liquid flows smoothly over a solid surface, the layer in contact with the solid tends to adhere, while the higher layers flow relative to it with different velocities (as in Fig. 16.7), where the velocity v increases with height y at a rate dv/dy, called the velocity gradient. Then, at any level, the liquid

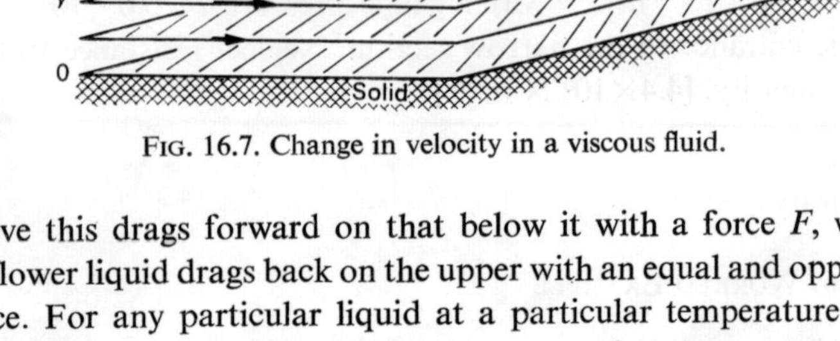

FIG. 16.7. Change in velocity in a viscous fluid.

above this drags forward on that below it with a force F, while the lower liquid drags back on the upper with an equal and opposite force. For any particular liquid at a particular temperature it is found that $F \propto A\, dv/dy$, where A is the area of the layer over which F is considered.

$$\therefore F = \eta A\, dv/dy, \tag{16.10}$$

where η is a constant of proportionality representing the "dynamic" viscosity of the particular liquid at the particular temperature. η usually decreases with rise of temperature. For water, $\eta \simeq 10^{-3}$ kg m^{-1} s^{-1} at 20°C, and decreases by about 3% per deg C near this temperature.

In the U-tube viscometer shown in Fig. 16.8, a specified volume of liquid is introduced into *D* and raised by suction into *C* until the level is above the constriction *A*. It is then allowed to run

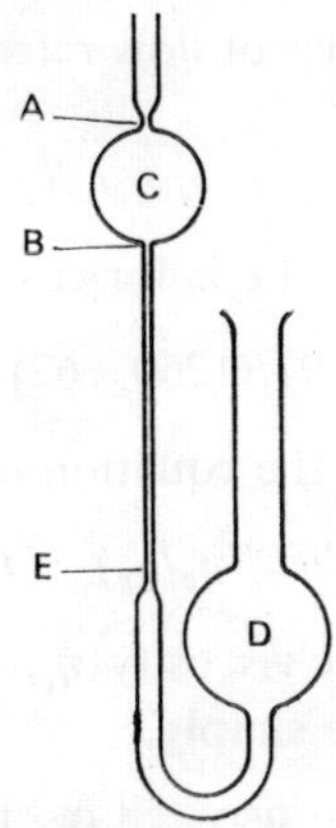

FIG. 16.8. U-tube viscometer (ex. 16.16).

back freely, while the level is timed from *A* to *B*. The velocity of the liquid in the fine bore "capillary" tube *BE* varies from a maximum down the axis to practically zero along the inner surface where the liquid tends to cling. Thus there is a velocity gradient radially across the tube and therefore viscous resistance to flow along the tube.

The velocity gradient at any point is proportional to the rate of flow, i.e. $dv/dr \propto \delta V/t$, where t is the time taken for the flow of a small volume δV. Since the flow occurs very slowly, with negligible acceleration of the liquid in the tube, the viscous drag δF on any small part of this liquid equals the driving force due to gravity and proportional to the density ϱ. Therefore $\delta F \propto \varrho$. But, in accordance with eqn. (16.10), $\eta \propto \delta F/(dv/dr)$.

$$\therefore\ \eta \propto \varrho t/\delta V. \tag{16.11}$$

This shows that the time taken for the flow of a liquid under gravity is proportional to η/ϱ, which is called the kinematic viscosity of the particular liquid at the particular temperature.

Solution

(a) Since the given times of flow refer to equal volumes, (16.11) gives $\eta \propto \varrho t$.

$$\therefore\ \eta_c = \eta_g \varrho_c t_c/(\varrho_g t_g),$$

where g is for glycerine and c is for castor oil.

$$\therefore\ \eta_c = 0.83\times 960\times 97/(1260\times 62) = 0.99\ \text{kg m}^{-1}\ \text{s}^{-1}. \qquad \text{(a)}$$

(b) Rearrangement of the equation used in (a) gives

$$(\eta_c/\varrho_c)/(\eta_g/\varrho_g) = t_c/t_g.$$

Therefore the kinematic viscosity (η_c/ϱ_c) may be calculated from the original data, or more simply,

$$\eta_c/\varrho_c = 0.99/960 = 1.0\times 10^{-3}\ \text{m}^2\ \text{s}^{-1}. \qquad \text{(b)}$$

16.17. Worked Example

A steel ball of diameter 2.0 mm falls at a constant rate of 10 cm in 3.4 s through oil in a relatively large jar. Calculate the dynamic viscosity of the oil at the temperature of the test, assuming the densities of the oil and the ball to be 920 kg m^{-3} and 7700 kg m^{-3} respectively.

Introduction

As the ball falls, the liquid must make way and allow the ball to slide through with some liquid carried on its surface. This results in a viscous drag on the ball, independent of the size of the jar if this is large enough not to restrict the way in which the liquid adjusts itself to the movement of the ball. Then the viscous drag is given by Stokes's formula $6\pi\eta rv$, where r is the radius of the ball, and v is the velocity through the liquid of viscosity η. This drag increases as the ball accelerates until it just equals the difference between the weight of the ball and the Archimedean upthrust due to the displacement of liquid (see ex. 8.7). Then, for constant velocity,

$$6\pi\eta rv = \tfrac{4}{3}\pi r^3 g(\varrho_1 - \varrho_2), \tag{16.12}$$

where $\frac{4}{3}\pi r^3$ is the volume of the ball, and ϱ_1 and ϱ_2 are the densities of the ball and liquid respectively.

Solution

Equation (16.12) simplifies to

$$\begin{aligned}\eta &= \tfrac{2}{9} r^2 g(\varrho_1 - \varrho_2)/v \\ &= (\tfrac{2}{9})\times 10^{-6}\times 9.81\times(7700-920)/(0.10/3.4) \\ &= 0.50 \text{ kg m}^{-1}\text{ s}^{-1}.\end{aligned}$$

16.18. Worked Example

A horizontal capillary tube of length 25 cm and internal diameter 0.50 mm is open at one end and is supplied at the other end with water at a pressure of 12 cmH_2O. Calculate the mean velocity of the water in the tube, assuming the viscosity of the water to be 1.3×10^{-3} kg m^{-1} s^{-1}.

Introduction

In steady streamline flow the velocity at each point in the tube is constant but decreases with radius from the axis, as shown approximately by the velocity vectors in Fig. 16.9. The velocity is practically zero at the inner surface of the tube where the liquid tends to cling.

Bernoulli's equation (16.8) shows that when the velocities v_1 and v_2 along the same streamline are equal at equal levels y_1 and y_2 at the ends of the tube, no pressure difference would be required if resistance to flow were negligible. In the present case, however, resistance occurs due to viscous drag of each cylindrical layer on its inner neighbour, and so a pressure difference Δp between the ends of the tube is required to maintain a constant rate of flow. If the tube is of very small cross-section ("capillary" tube) even the

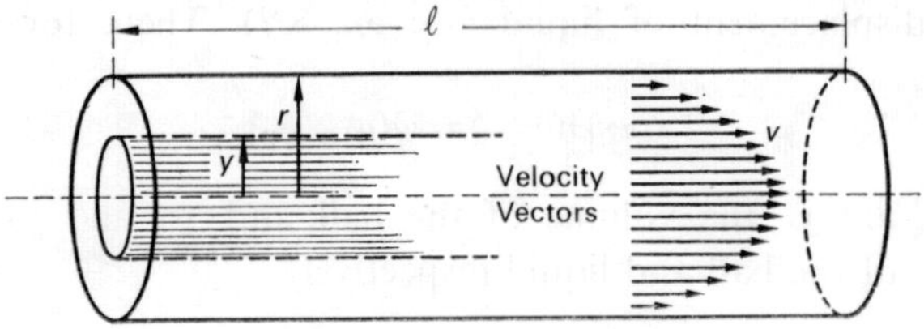

FIG. 16.9. Variation of velocity of a liquid in a horizontal tube.

axial velocity may be so low that the acceleration of the liquid as it enters the tube may be negligible. Then Δp is required only on account of viscosity.

Consider a cylindrical element of liquid, radius y, being the core of the liquid filling the tube of radius r in Fig. 16.9. The driving force on this element $= \Delta p \pi y^2$, and the opposing force due to viscosity acting over the cylindrical surface $= \eta(2\pi yl)(-dv/dy)$, where v is the velocity of flow and dv/dy is the velocity gradient [see eqn. (16.10)]. The negative sign is required since v decreases with increase of y as shown by the velocity vectors of Fig. 16.9.

Equating for zero acceleration of the element and integrating gives

$$(\Delta p/2l\eta) \int -y\,dy = \int dv.$$

$$\therefore\ (\Delta p/2l\eta)(-y^2/2) = v+k,$$

where k is a constant determined by putting $v = 0$ when $y = r$.

Thus $$k = -\Delta pr^2/4l\eta.$$

$$\therefore\ v = (\Delta p/4l\eta)(r^2-y^2). \tag{16.13}$$

Now the rate of volume flow through an annulus of radius y and radial thickness δy is $2\pi y\, \delta y v$. Therefore the total rate of volume flow

$$= \int_0^r 2\pi y\, dy\, \Delta p(r^2 - y^2)/4l\eta$$

$$= (\pi\, \Delta p/2l\eta)(r^4/2 - r^4/4) = \pi\, \Delta p r^4/(8l\eta). \quad (16.14)$$

Then the mean velocity $\bar{v}$ = (rate of volume flow)/(cross-section area)

$$= \Delta p r^2/(8l\eta). \quad (16.15)$$

Solution

The pressure difference

$$\Delta p = h\varrho g = 0.12 \times 10^3 \times 9.81 = 1.18 \times 10^3 \text{ N m}^{-2}.$$

Then eqn. (16.15) gives

$$\bar{v} = 1.18 \times 10^3 \times (0.25 \times 10^{-3})^2/(8 \times 0.25 \times 1.3 \times 10^{-3})$$
$$= 0.028 \text{ m s}^{-1}.$$

16.19. Calculate the pressure required to drive 1.0 l. of linseed oil per minute through a horizontal tube of internal diameter 4.0 mm and length 5.0 m, assuming a viscosity of 3.5×10^{-2} kg m^{-1} s^{-1}. [4.6×10^5 N m^{-2}.]

16.20. A small tank has an outlet in the base from which, leading vertically downward, is a capillary tube of length 10 cm and internal diameter 0.50 mm. The tank is filled to a depth of 5.0 cm with a liquid of density 870 kg m^{-3}. A volume of 1.6 cm^3 of the liquid is collected in 2.0 min. Calculate the viscosity. [1.5×10^{-3} kg m^{-1} s^{-1}.]

16.21. A droplet of oil mist of relative density 0.90 falls in air which has a viscosity of 1.8×10^{-5} kg m^{-1} s^{-1}. The terminal velocity is 5.0 cm s^{-1}. Calculate the weight of the droplet. [3.6×10^{-10} N.]

16.22. A horizontal tube 10 cm in length and 0.20 mm in diameter is sealed into the base of a larger vertical tube of 2.0 cm diameter having its axis vertical. A liquid of relative density 0.80 and viscosity 5.0×10^{-4} kg m^{-1} s^{-1} fills the tubes to a height of 10 cm above the outlet. (a) Calculate the initial rate at which the liquid level falls. Also, considering that the level falls exponentially with a time constant equal to the ratio of the initial height to the initial rate of fall, (b) calculate the time taken for the tank to become half empty. [(a) 2.0×10^{-6} m s^{-1}; (b) 9.8 h.]

CHAPTER 17

THERMODYNAMICS

Work by Expanding Gas

17.1. Worked Example

The p–V diagram for the cycle of a simple single-cylinder engine encloses an area of 30 cm² when pressure is represented on a scale of 2.5×10^5 N m^{-2} per cm, and volume on a scale of 15 cm³ per cm. The piston area is 20 cm², and the stroke is 9.0 cm. Calculate (a) the mean effective pressure, and (b) the indicated power at a speed of 250 cycles per minute.

Introduction

When gas in a cylinder expands and forces a piston outward, the pressure is to some extent unevenly distributed and varying in time. If, however, p is the average pressure over the piston area A at any instant, the force pA may be considered constant for a small interval of time during which the piston moves a slight distance δs, as in Fig. 17.1. Then the work δW done by the gas

$$= pA\,\delta s = p\,\delta V, \qquad (17.1)$$

where δV is the slight increase of volume of the gas (cf. ex. 16.2).

The work done in a larger expansion ΔV may be represented by a p–V diagram, as in Fig. 17.2, where the area under the graph represents the total work W, because the area equals the product of the average height which represents the average pressure p, and the base width which represents the expansion or swept volume ΔV. Usually there is some other pressure, possibly atmospheric, acting on the outside of the piston, in addition to a force in a connecting rod. Then the work transmitted by the connecting rod is

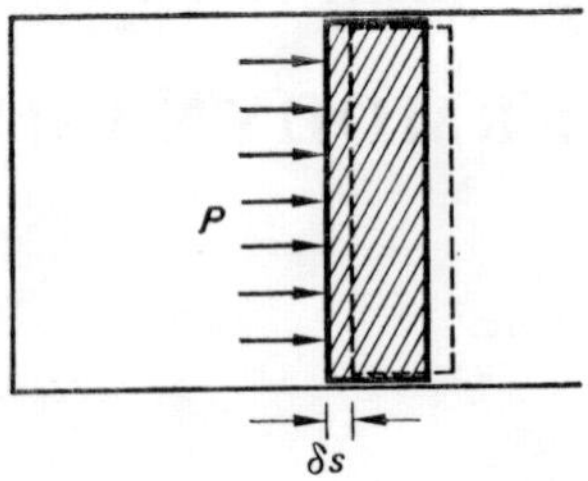

FIG. 17.1. Piston motion.

given by eqn. (17.1) and Fig. 17.2 if p is not the absolute internal pressure but the excess pressure on the inside face of the piston compared with that on the outside.

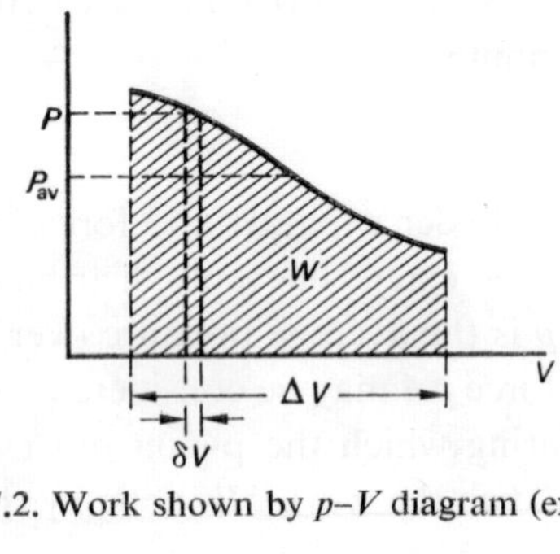

FIG. 17.2. Work shown by p–V diagram (ex. 17.1).

In a piston engine operating under constant conditions, a certain variation of pressure is repeated in cycles. Thus the p–V diagram of Fig. 17.3 shows a simplified petrol engine cycle. The useful work done by the expansion is shown by the area between the curve for the outward part of the movement and the straight line showing atmospheric pressure p_A, since the height of this area

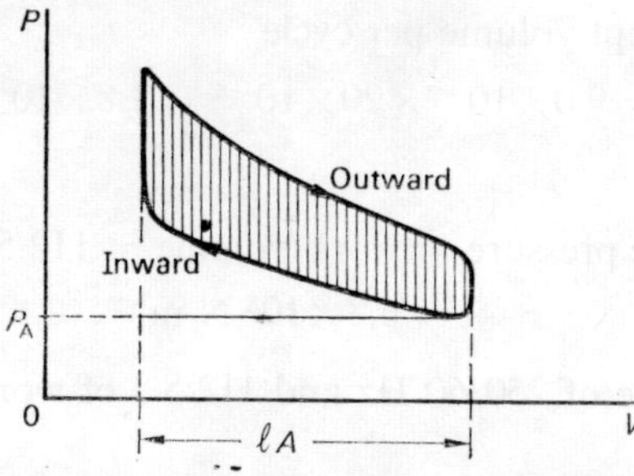

FIG. 17.3. Simplified petrol engine cycle.

represents internal excess pressure. After this expansion the spent gas is exhausted and replaced by fresh as the piston moves back and out again with changes of pressure which are ignored. Finally, as the piston returns, the pressure builds up in the cylinder causing some wastage of work shown by the area under the curve for the inward movement. Then the net work done per cycle is shown by the area enclosed by the loop of the p–V diagram.

The width of this loop represents the volume lA swept by the piston of area A during its stroke of length l. The mean height of the loop, i.e. the mean length of the vertical lines shown in the loop, represents the "mean effective pressure" during the cycle. Thus the enclosed area represents the work per cycle

$$= (\text{swept volume})\times(\text{mean effective pressure}). \qquad (17.2)$$

Solution

(a) Since a height of 1.0 cm on the p–V diagram represents a pressure of 2.5×10^5 N m^{-2} and a width of 1.0 cm represents a volume of 15 cm^3, the area of each square centimetre represents work of

$$2.5\times10^5\times15\times10^{-6} = 3.75 \text{ J}.$$

Therefore the loop area of 30 cm^2 represents

$$W = 30\times3.75 = 112.5 \text{ J}.$$

Since the swept volume per cycle

$$= lA = 9.0\times 10^{-2}\times 20\times 10^{-4} = 1.8\times 10^{-4}\ \text{m}^3,$$

then (17.2) gives

$$\text{mean effective pressure} = \text{work/volume} = 112.5/(1.8\times 10^{-4})$$
$$= 6.3\times 10^5\ \text{N m}^{-2}. \quad \text{(a)}$$

(b) For a rate of 250/60 Hz and 112.5 J of work per cycle, the indicated power

$$= 112.5\times 250/60 = 470\ \text{W}. \quad \text{(b)}$$

17.2. Worked Example

A mass of air occupying 1.0 l. at s.t.p. expands to a volume of 15.0 l., (a) at constant pressure, and (b) at constant temperature. Calculate the external work done in each case. Ref.: s.t.p.

Introduction

When a gas expands by ΔV at constant pressure P, the work done by the gas is given as in eqn. (17.1) by

$$W = p\,\Delta V = p(V_2 - V_1), \quad (17.3)$$

where V_1 and V_2 are the initial and final volumes respectively [see Fig. 17.4(a)].

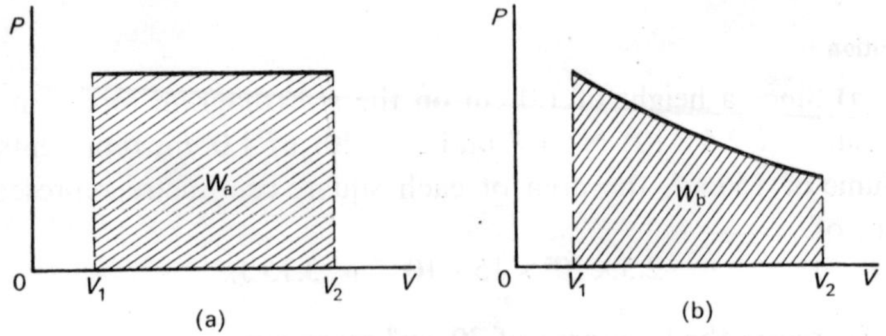

FIG. 17.4. (a) Expansion at constant pressure. (b) Expansion at varying pressure.

If the pressure varies during the expansion, as shown for example in Fig. 17.4(b), the work done may be different and indicated by a different area under the p–V line. If p is a simple function of V, W may be calculated by integration:

$$W = \int p\,dV. \tag{17.4}$$

Thus, for constant temperature (isothermal) changes in which the pressure is uniformly distributed throughout the gas, Boyle's law [eqn. (9.1)] gives $p = k/V$, where k is a constant. Then

$$W = \int_{V_1}^{V_2} p\,dV = \int_{V_1}^{V_2} (k/V)\,dV = \Big[k \log_e V\Big]_{V_1}^{V_2}$$

$$= k(\log_e V_2 - \log_e V_1) = pV \log_e (V_2/V_1), \tag{17.5}$$

where pV is the constant product for the particular temperature.

Solution

(a) For expansion from 1.0×10^{-3} m^3 to 15×10^{-3} m^3 at constant pressure of 1.01×10^5 N m^{-2}, eqn. (17.3) gives

$$W = 1.01\times10^5(15-1)\times10^{-3} = 1.4\times10^3 \text{ J.} \tag{a}$$

(b) For isothermal expansion, $pV = 1.01\times10^5\times1.0\times10^{-3}$ J. Then eqn. (17.5) gives

$$W = 1.01\times10^2 \log_e \left(\tfrac{15}{1}\right) = 101\times2.3\times1.176 = 270 \text{ J.} \tag{b}$$

17.3. Calculate the work done by the expansion of 200 g of gas from a density of 0.50 kg m^{-3} to 0.25 kg m^{-3} at a constant pressure 1.0×10^5 N m^{-2}. [4.0×10^4 J.]

17.4. A hot-air central heating system takes a volume of 5.0×10^3 m^3 of air per hour and raises its temperature from 7.0°C to 17.0°C at constant standard atmospheric pressure. Calculate the work done per hour in causing expansion of the air. [1.8×10^7 J.]

17.5. Calculate the external work done when 5.0 g of carbon dioxide is heated through 50 deg C at constant pressure. Assume: relative molecular mass of $CO_2 = 44$. Ref.: gas constant. [47 J.]

17.6. Calculate (a) the mean pressure, and (b) the work done when 1.0 kilomole of ideal gas expands from 1.0 m^3 to 2.0 m^3 isothermally at a temperature of 0°C with uniform distribution of pressure. Check the calculation by determining the area under the *p–V* graph. Ref.: ideal gas volume. [(a) 1.6×10^6 N m^{-2}; (b) 1.6×10^6 J.]

Heat, Work, and Internal Energy

17.7. Worked Example

Calculate (a) the increase of internal energy, and (b) the external work done, during the boiling of 5.0 kg of water to saturated steam at constant temperature of 200°C and pressure of 1.6×10^6 N m^{-2}. The latent heat at 200°C is 1.96×10^6 J kg^{-1} and the densities of saturated steam and water at 200°C are respectively 7.8 kg m^{-3} and 860 kg m^{-3}.

Introduction

In the question the water which changes from liquid to steam forms what may be called a system, as distinct from the surroundings which are the container and everything outside this. All energy exchanged between the system and the surroundings may be divided into two categories, heat and work. Also the system may store "internal energy" in various forms, such as potential energy due to attraction between particles and kinetic energy due to random thermal movements.

If the system takes in heat Q from the surroundings while expanding and doing work W on the surroundings, any difference between this input and output of energy must be accounted for by a change ΔU of the internal energy of the system. Thus

$$Q = \Delta U + W \tag{17.6}$$

if all quantities are measured in the same units (joules). This is the first law of thermodynamics, and is a statement of the conservation of energy in relation to the exchanges of heat and work between a system and its surroundings. Q would be negative if the system lost heat; W would be negative if work were done on the system; ΔU would be negative if the internal energy decreased.

Note that heat and work are not always two different forms of energy but are quantities of energy transferred under different conditions. Thus in the particular case, heat may be conducted into the system (+ve Q) by a transfer of kinetic energy from molecules of the surroundings at a higher temperature in collisions with molecules of the system at a lower temperature. Also work (−ve W) may be done by similar collisions in which again kinetic energy is gained by molecules of the system when the boundary of the system moves inward. The distinction lies in the different conditions, of temperature difference in the first case and boundary movement in the second. In the case of heat flow, the molecules at the higher temperature have a higher average thermal kinetic energy and so increase the thermal motion of the less energetic molecules at the lower temperature. In the case of work done on the gas by compression, gas molecules gain kinetic energy as they rebound from the other molecules across the boundary when the boundary advances against them, somewhat like an ideal bat moving against an ideal ball.

In calculating the work from the gas pressure and volume changes it is assumed in all the following problems that the pressure is uniform throughout the volume at all instants. Then calculations can be based on eqns. (17.3) and (17.4). Thus in this problem it is assumed that the boundary of the system expands as steam is produced, but at a rate which is negligible in comparison with the velocities of the molecules of steam. Otherwise the pressure on the boundary would be reduced, tending to zero in the extreme when the molecules of steam would fail to keep up with the receding boundary.

Solution

The heat supplied at constant temperature

$$(\text{mass})\times(\text{latent heat}) = 5.0\times 1.96\times 10^6 = 9.8\times 10^6 \text{ J}.$$

The work done at constant pressure $= p(V_s - V_w)$, where $V_s =$ volume of steam $= 5.0/7.8 = 0.64 \text{ m}^3$ and $V_w =$ volume of water $= 5.0/860 = 5.8\times 10^{-3} \text{ m}^3$.

$$W = 1.6\times 10^6\times(640-5.8)\times 10^{-3} = 1.0\times 10^6 \text{ J}. \qquad \text{(b)}$$

Then, from eqn. (17.6),

$$\Delta U = Q - W = (9.8-1.0)\times 10^6 = 8.8\times 10^6 \text{ J}. \qquad \text{(a)}$$

17.8. Calculate the increase of internal energy of 100 g of water when converted from liquid at 0°C to steam at constant pressure of 7.0×10^5 N m^{-2}. At this pressure, the boiling point is 165°C, the latent heat is 2.1×10^6 J kg^{-1}, the specific volume of water is 1.1×10^{-3} m^3 kg^{-1}, and the specific volume of steam is 0.27 m^3 kg^{-1}. The mean specific heat of water in this range is 4.2×10^3 J kg^{-1} deg C^{-1}. [2.6×10^5 J.]

17.9. Given the latent heat of vaporization of water to be 2.0×10^6 J kg^{-1} at a pressure of 1.4×10^6 N m^{-2}, calculate the quantity of heat which must be supplied to obtain 2.0×10^6 J of external work by converting water to steam at this pressure throughout. Density of steam $= 7.0$ kg m^{-3}, and of water $= 870$ kg m^{-3}. [2.0×10^7 J.]

17.10. A certain mass of gas has an initial volume of 0.60 l. at s.t.p. Assuming that the internal energy of the gas depends only upon its temperature, calculate the heat which must be removed from the gas to maintain constant temperature while undergoing uniform compression to a volume of (a) 0.40 l., and (b) 0.20 l. [(a) 25 J; (b) 67 J.]

17.11. Assuming that the internal energy of an ideal monatomic gas is entirely due to kinetic energy of translation, calculate the heat required to raise the temperature of 1.0 kilomole of such a gas by 20 deg C, while the volume is maintained constant. Ref.: gas constant. [2.5×10^5 J.]

Principal Specific Heats of Gases

17.12. Worked Example

Calculate the temperature rise when 1000 J of heat is taken by a mass of 100 g of oxygen maintained at constant pressure. Specific heat of oxygen at constant volume = 653 J kg^{-1} deg C^{-1}. Relative molecular mass of oxygen = 32. Ref.: universal gas constant.

Introduction

When gas is heated, the rise of temperature depends not only upon the type and mass of gas but also upon the use which is made of the heat energy supplied. Thus if the gas is heated at constant volume, the energy supplied as heat Q is retained internally as ΔU, but if expansion is allowed, some energy is lost by the work done and less is retained to cause a temperature rise. An infinite variety of conditions may occur, but two principal conditions are considered:

(1) At constant volume,

$$Q_1 = \Delta U = mc_v \Delta T, \tag{17.7}$$

where Q_1 is the heat required to raise the temperature of a mass m of gas by ΔT when c_v is the specific heat at constant volume.

(2) At constant pressure,

$$Q_2 = mc_p \Delta T, \tag{17.8}$$

where c_p is the specific heat at constant pressure.

If ΔT and ΔU are the same in cases (1) and (2), the difference between Q_1 and Q_2 is due solely to external work done in case (2).

$$\therefore Q_2 = Q_1 + p\,\Delta V,$$

where $p\,\Delta V$ is the work done.

$$\therefore mc_p\,\Delta T = mc_v\,\Delta T + p\,\Delta V. \tag{17.9}$$

Now from eqn. (15.5), $pV = (mR/M_r)T$.

$\therefore$ at constant pressure, $p\,\Delta V = (mR/M_r)\,\Delta T$.

$\therefore$ eqn. (17.9) becomes $mc_p\,\Delta T = mc_v\,\Delta T + (mR/M_r)\,\Delta T$.

$$\therefore c_p - c_v = R/M_r. \tag{17.10}$$

For an ideal gas c_v is independent of pressure, and so then is c_p.

Solution

Since $R = 8.31 \times 10^3$ J K^{-1} per kilomole, and $M_r = 32$, $R/M_r = 259$ J kg^{-1} K^{-1}. Since the magnitude of 1 K is the same as that of 1 deg C, these units may be used together in eqn. (17.10).

$$\therefore c_p = 653 + 259 = 912 \text{ J kg}^{-1}\text{ K}^{-1}.$$

Then, from eqn. (17.8),

$$\Delta T = Q/(mc_p)$$

$$= 1000/(0.100 \times 912) = 11 \text{ K or } 11 \text{ deg C}.$$

17.13. Worked Example

The temperature of a mass of 1.8 g of hydrogen is changed from 20°C to 100°C, and its volume from 5.0 l. to 50 l. Calculate the net amount of heat required and the external work done (a) when the gas is heated from 20°C to 100°C at constant volume and then expanded isothermally to the final volume, and (b) when the gas is cooled at constant volume to the final pressure and then heated at constant pressure to the final volume. Assume the density at s.t.p. to be 0.090 kg m^{-3}, the specific heat at constant volume to be 1.02×10^4 J kg^{-1} deg C^{-1}, and the specific heat at constant pressure to be 1.44×10^4 J kg^{-1} deg C^{-1}.

Introduction

The two processes (a) and (b) are seen to involve different amounts of work if the p–V diagrams are drawn as in Figs. 17.5(a) and 17.5(b). The areas under the p–V lines represent the work done and show that the change from the initial point (i) to the final point

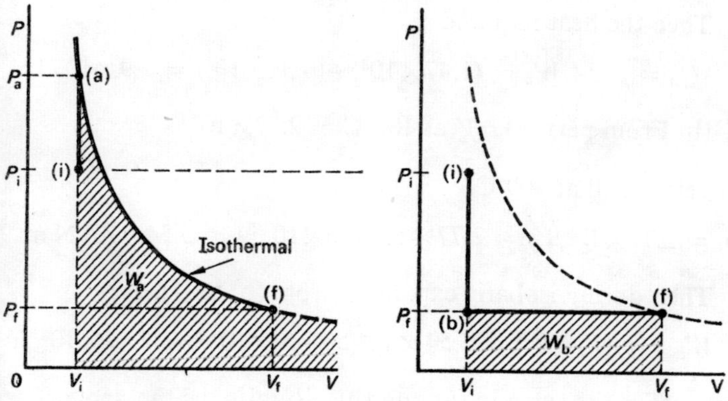

(a) (b)

FIG. 17.5. (a) Ex. 17.13(a). (b) Ex. 17.13(b).

(f) gives more work W_a in expanding via point (a) than W_b via point (b). The calculation assumes slow expansions and rates of heating in order to maintain uniform distributions of p and T throughout (see also ex. 17.7).

Solution

(a) During heating from (i) to (a) at constant V_i, the heat is taken as internal energy.

$$\therefore \; \Delta U = mc_v \Delta T = 1.8\times10^{-3}\times1.02\times10^{4}\times(100-20)$$
$$= 1.47\times10^{3} \text{ J}.$$

Then, during isothermal expansion from (a) to (f), eqn. (17.5) gives

$$W_a = pV \log_e (V_f/V_i),$$

where $V_f/V_i = \frac{50}{5} = 10$, and $pV = pm/\varrho$, where $m = 1.8\times10^{-3}$ kg, p = standard pressure $= 1.013\times10^5$ N m^{-2}, and ϱ = density at (273+100) K and same p, $= 0.090\times273/373$ kg m^{-3}.

$$\therefore\; W_a = (1.013\times10^5\times1.8\times10^{-3}/0.090)(373/273)\log_e 10$$
$$= (2.77\times10^3)\times2.303 = 6.38\times10^3 \text{ J}.$$

Then the heat required

$$Q_a = \Delta U + W_a = (1.47\times10^3)+(6.38\times10^3) = 7.9\times10^3 \text{ J}.$$

(b) From part (a), pV at 100°C $= 2.77\times10^3$.

Therefore still at 100°C,

$$p_f = 2.77\times10^3/V_f = 2.77\times10^3/(50\times10^{-3}) = 5.54\times10^4 \text{ N m}^{-2}.$$

Then during isobaric expansion from (b) to (f),

$$W_b = p_f(V_f - V_i) = 5.54\times10^4(50-5)\times10^{-3} = 2.49\times10^3 \text{ J}.$$

Since ΔT is the same in (a) and (b), ΔU also is the same. Therefore $\Delta U = 1.47\times10^3$ J, and the heat required is

$$Q_b = \Delta U + W_b = 1.47\times10^3 + 2.49\times10^3 = 4.0\times10^3 \text{ J}. \quad \text{(b}$$

17.14. Nitrogen flows continuously at a rate of 3.0 cm³ s⁻ into a tube in which it is heated by an electric element carrying current of 0.16 A at a p.d. of 2.1 V. The temperature rises fro 10°C at the inlet to 90°C at the outlet while the pressure remai at 760 mmHg. Assuming negligible heat losses, calculate th specific heat of nitrogen at constant pressure, taking the densi to be 1.25 kg m^{-3} at s.t.p. [1.1×10^3 J kg^{-1} deg C^{-1}.]

17.15. Assuming the density of air to be 1.29 kg m^{-3} at s.t. and the specific heat to be 1000 J kg^{-1} deg C^{-1} at constant pressur calculate the specific heat at constant volume. [713 J kg^{-1} deg C^{-1}

17.16. A quantity of carbon dioxide occupies a volume 2.0×10^{-3} m³ at s.t.p. Calculate the net quantity of heat requir by the gas and the net external work done on the gas (a) if it first heated to 75°C at constant volume, and then compressed at constant temperature to a pressure of 1520 mmHg, and (b) if it is first compressed at constant temperature to a pressure of 1520 mmHg and then heated at constant pressure to a temperature of 75°C. Assume a density of 1.98 kg m^{-3} at s.t.p., and specific heats $c_p = 850$ and $c_v = 650$ J kg^{-1} deg C^{-1}. [(a) 77 J, 116 J; (b) 112 J, 85 J.]

Adiabatic Changes

17.17. Worked Example

The temperature of a mass of 10 g of hydrogen is decreased from +20°C to −20°C by adiabatic expansion. Calculate the loss of internal energy assuming $\gamma = 1.41$, $M_r = 2$, and $R = 8310$ J K^{-1} per kilomole.

Introduction

An adiabatic process is one in which no energy is taken in or given out as heat Q, although work may be done on or by the system during the process. Thus if a gas expands adiabatically, $Q = 0$, and any work done by the gas involves a drain on its internal energy U with a consequent fall of temperature. Conversely, the temperature rises during adiabatic compression since U increases if no heat may escape while work is done on the gas. In both cases eqn. (17.6) becomes $Q = 0 = \Delta U + W$, and eqn. (17.7) gives $\Delta U = mc_v \Delta T$. Therefore the work done by a gas during an adiabatic change,

$$W = -\Delta U = -mc_v \Delta T. \qquad (17.11)$$

Rapid fluctuations of pressure and density, as in a sound wave, are practically adiabatic and result in fluctuations of temperature also, since little time is available for heat flow before temperature differences disappear or are reversed.

When p and T are uniform throughout V, the product pV for an ideal gas changes with T as in eqn. (15.5).

$$\therefore\ \Delta(pV) = (Rm/M_r)\,\Delta T.$$

But, from eqn. (17.10),

$$R/M_r = c_p - c_v.$$

$$\therefore\ \Delta(pV) = m(c_p - c_v)\,\Delta T.$$

Then, substituting for ΔT in eqn. (17.11) and using the symbol γ for the ratio c_p/c_v gives

$$W = -mc_v\,\Delta(pV)/(m(c_p - c_v)) = \Delta(pV)/(1-\gamma) \qquad (17.12)$$

for an adiabatic change in which p is uniform throughout V.

Note that for the ideal gas of the kinetic theory of ex. 15.21, $U = 3pV/2$. Then for adiabatic conditions,

$$W = -\Delta U = -3\Delta(pV)/2.$$

Comparison with eqn. (17.12) gives $\gamma = 1\frac{2}{3}$ for this ideal gas, as found experimentally for argon and helium. For polyatomic gases γ is found to be less than $1\frac{2}{3}$ and the change of (pV) and T correspondingly less for the same W, when $Q = 0$, due to additional forms of internal energy such as rotational kinetic energy of large molecules.

Solution

First, from eqn. (15.5),

$$\Delta(pV) = (Rm/M_r)\,\Delta T = 8310\times 10\times 10^{-3}\times(-40)/2 = -1662.$$

Then, from eqn. (17.12),

$$W = \Delta(pV)/(1-\gamma) = -1662/(-0.41) = 4.1\times 10^3\ \text{J}.$$

Therefore, from eqn. (17.11), $\quad -\Delta U = 4.1\times 10^3$ J.

Alternatively, c_v is found first from eqn. (17.10): $c_p - c_v = R/M_r$

$$\therefore\ c_v = R/(M_r(\gamma-1)) = 8310/(2\times 0.41) = 1.014\times 10^4.$$

Then, from eqn. (17.11),

$$\begin{aligned} -\Delta U &= -mc_v\,\Delta T \\ &= -(10\times 10^{-3})\times 1.014\times 10^4\times(-40) = 4.1\times 10^3\ \text{J}. \end{aligned}$$

17.18. Worked Example

A constant mass of gas at a temperature of 50°C and a pressure of 350 mmHg is slowly compressed to one-tenth of its initial volume. Calculate the pressure and temperature developed if no heat is lost or gained. $\gamma = 1.4$.

Introduction

During an adiabatic expansion or compression of an ideal gas in which p and T are uniform throughout V, p and V are related not only by the gas equation (15.5), i.e. $pV/T = mR/M_r$, but also by an equation related particularly to the adiabatic condition. Figure 17.8 (p. 287) shows roughly a curve ba of this equation, crossing isothermals such as bc for one temperature and da for a lower temperature because the temperature falls in adiabatic expansion and rises in adiabatic compression.

The equation may be obtained by first combining eqns. (17.3) and (17.12):

$$W = \Delta(pV)/(1-\gamma) = p\,\Delta V.$$

Then, considering an infinitesimal change, dividing by pV, and integrating, gives

$$\int d(pV)/(pV) = (1-\gamma)\int dV/V.$$

$$\therefore\ \log_e (pV) = (1-\gamma)\log_e V + \text{a constant},$$

$$\therefore\ \log_e (pV) - \log_e V^{1-\gamma} = \text{a constant},$$

$$\therefore\ pV/V^{1-\gamma} = pV^{\gamma} = \text{a constant}, \qquad (17.13)$$

for an adiabatic change in which p is uniform throughout V.

Solution

For an adiabatic change from condition 1 to condition 2, eqn. (17.13) gives $p_1V_1^{\gamma} = p_2V_2^{\gamma}$, where p_1 and p_2 may be in any units provided they are the same, and likewise V_1 and V_2.

Then $$p_2 = p_1(V_1/V_2)^{\gamma} = 350 \times 10^{1.4}$$
$$= 350 \times 25.1 = 8800 \text{ mmHg}.$$

Since the ideal gas equation applies to any change of constant mass of ideal gas,

$$p_1V_1/T_1 = p_2V_2/T_2.$$

$$\therefore\ T_2 = T_1p_2V_2/(p_1V_1) = 323 \times 8800/(350 \times 10)$$
$$= 812 \text{ K} = 540°\text{C} \quad \text{(2 sig. fig.)}.$$

17.19. The volume of a constant mass of gas having $\gamma = 1.4$ is changed adiabatically to (a) half and (b) double the initial value. Calculate the change of pressure in each case. [(a) +160%; (b) −62%.]

17.20. A quantity of carbon dioxide, initially occupying a volume of 0.20 m³ at a temperature of 25°C and pressure of 1000 mmHg, expands adiabatically till the temperature reaches 5°C. Calculate the external work done given the specific heat at constant volume to be 650 J kg^{-1} deg C^{-1} and the density at s.t.p. to be 2.0 kg m^{-3}. [6.3×10^3 J.]

17.21. A quantity of methane having an initial volume of 0.50 m³ at 0°C expands from a pressure of 2.0 atm to 0.50 atm, adiabatically. Calculate (a) the final volume, (b) the final temperature, (c) the external work done, and (d) the mean pressure during expansion. $\gamma = 1.31$ [(a) 1.4 m³; (b) 200 K; (c) 9.1×10^4 J; (d) 0.97×10^5 N m^{-2}.]

17.22. 1.0 g of hydrogen at 0°C is compressed adiabatically to a quarter of its original volume. Calculate (a) the work required, (b) the factor by which the pressure increases, and (c) the final temperature. Assume $\gamma = 1.41$, $M_r = 2$. Ref.: *R*. [(a) 2.1×10^3 J; (b) 7.1; (c) 480 K.]

17.23. A vessel containing air at 15°C and 1000 mmHg is opened and closed again so that the pressure is reduced to 760 mmHg adiabatically. The pressure then rises gradually to 820 mmHg as inflow of heat restores the temperature to 15°C. Calculate (a) the lowest temperature reached, and (b) the ratio of principal specific heats of air. [(a) −6°C; (b) 1.4.]

Reversible Cycles

17.24. Worked Example

A heat engine which is to provide an output power of 100 kW is to work between a heat source at a temperature of 200°C and a heat sink at 20°C. Calculate (a) the maximum possible efficiency, and (b) the minimum possible rate of discharge of heat to the sink, when the engine works at full power.

Introduction

The common types of reciprocating heat engine are devices for directing some of the energy of random thermal motion of gas molecules into useful work when a piston is driven by the gas expanding after combustion in a cylinder. During each complete cycle of events in the cylinder, not only is heat produced by combustion at high pressure, but some heat is taken out by the exhaust and cooling system in order to regain the starting point for the next cycle. Similarly, not only is work done by an expansion stroke, but some work is taken back during a compression stroke.

Then, if in each cycle Q_1 is the heat provided by combustion, Q_2 is the heat rejected, and W is the resultant output of work, the efficiency η is defined as the proportion of Q_1 which is converted to W, i.e. $\eta = W/Q_1$. But because energy cannot be created or destroyed, $Q_1 = Q_2 + W$ when Q and W are measured in the same units.

$$\therefore\ \eta = W/Q_1 = (Q_1 - Q_2)/Q_1 = 1 - (Q_2/Q_1). \qquad (17.14)$$

According to Carnot's theorem, the maximum value of this efficiency is limited by the highest and lowest temperatures T_1 and T_2, and the maximum value would be obtained if all Q_1 were taken at T_1 and all Q_2 were rejected at T_2 in such ways that no irreversible changes occurred. One factor which affects reversibility is the speed of operation, since gas pressure and density tend to build up in front of a rapidly approaching surface, but to fall behind a receding one (see also ex. 17.7). Thus if the rapid movement of a piston were reversed, the change of pressure would not be exactly reversed in the same time.

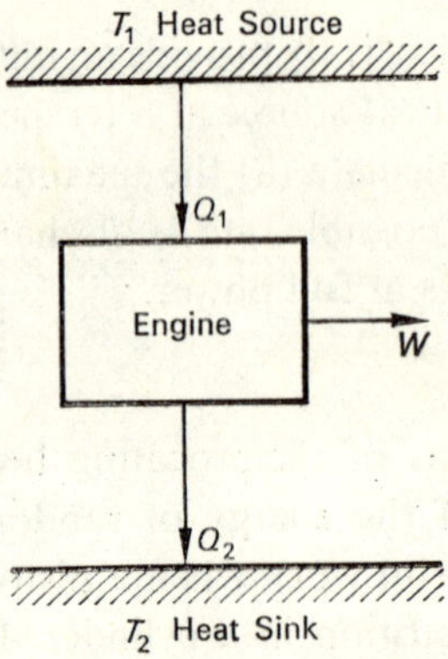

FIG. 17.6. Heat transfer between source and sink—heat engine.

Figure 17.6 shows in outline an imaginary arrangement of a "source" and "sink" of heat at constant temperatures and an engine which automatically regulates the flow of heat in and out of a constant mass of gas which it contains. Q_1, Q_2, and W refer to one cycle of expansion and contraction.

The relationship of η_{max} to T_1 and T_2 may be obtained by considering the gas to be ideal and to take in an infinitesimal amount of heat dQ at temperature T. Then, assuming uniform distribution of pressure and temperature, eqs. (17.1), (17.6), (17.7), and (15.5) apply. Thus for the infinitesimal quantities,

$$\begin{aligned} dQ &= dU+dW = dU+p\,dV \\ &= mc_v\,dT+mRT\,dV/(M_rV). \end{aligned}$$

Dividing by T and integrating gives

$$\begin{aligned} \Sigma(dQ/T) &= mc_v \int dT/T+(mR/M_r)\int dV/V \\ &= mc_v \log_e(T_f/T_i)+(mR/M_r)\log_e(V_f/V_i) \end{aligned}$$

for a change from an initial state i to a final state f. But for a complete cycle, i and f are the same. Therefore $T_i = T_f$ and $V_i = V_f$, making both logarithms zero.

$$\therefore \ \Sigma\, dQ/T = 0 \tag{17.15}$$

for a complete reversible cycle. Now in the cycle of maximum efficiency to which Fig. 17.6 refers, there are simply two values of Q at corresponding constant values of T. Therefore, reckoning Q_1 positively into the engine and Q_2 negatively out of it,

$$\Sigma dQ/T = Q_1/T_1 - Q_2/T_2 = 0. \qquad \therefore \ Q_2/Q_1 = T_2/T_1.$$

Then, from eqn. (17.14), the maximum possible efficiency

$$\eta_{max} = 1 - Q_2/Q_1 = 1 - T_2/T_1. \tag{17.16}$$

Solution

(a) Equation (17.16) gives

$$\eta_{max} = 1-(273+20)/(273+200)$$
$$= 1-0.62 = 0.38 = 38\%. \qquad \text{(a)}$$

(b) If f is the frequency in cycles per second, and in each cycle $Q_2 = Q_1 - W$, the rate of discharge of heat to the sink $= fQ_2 = fQ_1 - fW$. But $fQ_1 = fW/\eta_{max} = 100/0.38$ kW $= 260$ kW.

$\therefore$ minimum rate of discharge of heat $= 260 - 100 = 160$ kW. (b)

17.25. Worked Example

A refrigerator operating between two constant temperatures has a coefficient of performance of 6.0. Calculate the coefficient of performance when viewed as a heat pump, and calculate the lower temperature if the higher is 50°C and the cycle is perfectly reversible.

Introduction

A refrigerator or heat pump may be considered to be a heat engine operating in reversed cycles. Thus by supplying mechanical work to a suitable engine it may be driven as a compressor which raises the temperature of gas by compressing it and lowers the temperature by allowing it to expand. When hot at temperature T_1 the gas is allowed to give out heat Q_1, and when cold at T_2 it is allowed to take in Q_2. Figure 17.7 shows the reversal of Q_1, Q_2, and W to be compared with Fig. 17.6.

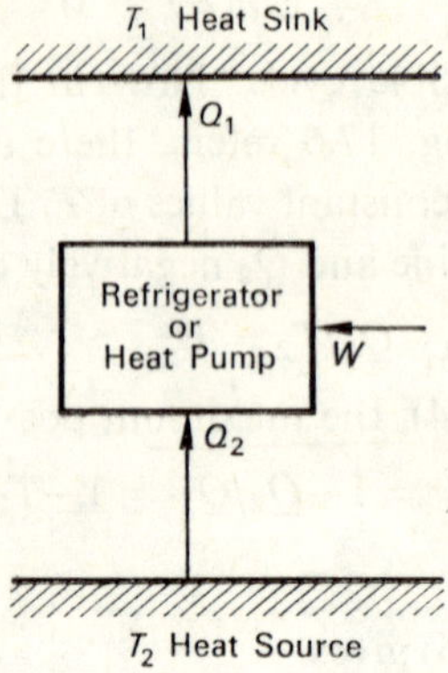

FIG. 17.7. Heat transfer between source and sink—heat pump.

The main distinction between the refrigerator and the heat pump lies in the purpose of the device. Thus the device is a refrigerator when the purpose is to maintain a low temperature T_2 by extracting Q_2, and is a heat pump when the purpose is to provide Q_1 at a higher temperature T_1. Both effects occur together in a domestic refrigerator which extracts heat from food and dissipates heat to its surroundings.

The coefficient of performance as a refrigerator epuals

$$Q_2/W \tag{17.17}$$

since the extraction of Q_2 is the desirable result of using W.

The coefficient of performance as a heat pump epuals

$$Q_1/W \tag{17.18}$$

since the output of Q_1 is the desirable result of using W in this case.

The ideal relationship between Q_1, Q_2, T_1, and T_2 is the same for both directions of operation of a perfectly reversible cycle, since eqn. (17.15) applies. Therefore in the case of an ideal refrigerator or heat pump, as for an ideal heat engine,

$$Q_2/Q_1 = T_2/T_1. \tag{17.19}$$

Solution

Since for the refrigerator $Q_2/W = 6$, for the heat pump

$$Q_1/W = (Q_2+W)/W = Q_2/W+1 = 6+1 = 7.$$

Then

$$Q_2/Q_1 = (Q_2/W)/(Q_1/W) = \frac{6}{7} = T_2/T_1$$

from eqn. (17.19). Therefore

$$T_2 = \frac{6}{7}\times(273+50) = 277 \text{ K} = 4°\text{C}.$$

17.26. Worked Example

A quantity of gas, initially at a temperature of 27°C, is first compressed adiabatically to a quarter of the initial volume, then expanded isothermally back to the initial volume, and finally cooled at constant volume back to the initial pressure. The ratio of principal specific heats of the gas = 1.40. Assuming the cycle to be perfectly reversible, compare its efficiency with the maximum efficiency possible for a cycle between the same maximum and minimum temperatures.

Solution

To find the resultant work done during the cycle, consider the three stages of the cycle shown in Fig. 17.8.

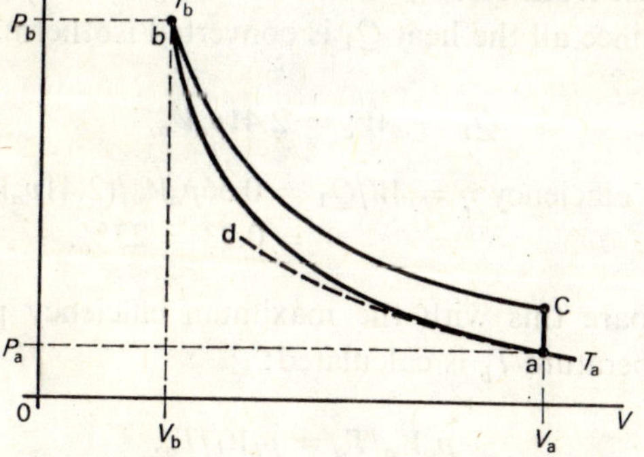

FIG. 17.8. Adiabatic curve between two isothermals (ex. 17.26).

Firstly, for the adiabatic compression from a to b,

$$p_a V_a^\gamma = p_b V_b^\gamma. \qquad \therefore\ p_b = p_a(V_a/V_b)^\gamma = p_a(4)^{1.4} = 6.96p_a.$$

Then, from eqn. (17.12)

$$\begin{aligned} {}_aW_b &= (p_bV_b - p_aV_a)/(1-\gamma) \\ &= (6.96/4-1)p_aV_a/(1-1.4) = -1.85p_aV_a. \end{aligned}$$

This is negative because work is done on the gas and not by the gas (see ex. 17.7).

Secondly, for the isothermal expansion from b to c, eqn. (17.5) gives

$$\begin{aligned} {}_bW_c &= p_bV_b \log_e (V_c/V_b) = (6.96/4)p_aV_a \log_e 4 \\ &= 1.74 \times 1.386 p_aV_a = 2.41p_aV_a. \end{aligned}$$

Finally, for the change from c to a at constant volume, no work is done.

Therefore, for the complete cycle the resultant work done by the gas equals

$${}_aW_b + {}_bW_c = -1.85p_aV_a + 2.41p_aV_a.$$

$$\therefore\ \text{resultant } W = 0.56p_aV_a.$$

A supply of heat Q_1 to the gas is required only from b to c to maintain constant temperature while work is being done, whereas heat Q_2 is lost from c to a, and no exchange of heat occurs from a to b. But since all the heat Q_1 is converted isothermally to work ${}_bW_c$,

$$Q_1 = {}_bW_c = 2.41p_aV_a.$$

$$\begin{aligned} \therefore\ \text{efficiency } \eta = W/Q_1 &= 0.56p_aV_a/(2.41p_aV_a) \\ &= 0.23 = 23\%. \end{aligned}$$

To compare this with the maximum efficiency possible, the highest temperature T_b is calculated:

$$p_aV_a/T_a = p_bV_b/T_b.$$

$$\therefore\ T_b = T_a p_bV_b/(p_aV_a) = 300 \times 6.96/4 = 522\ \text{K}.$$

Then, from eqn. (17.16),

$$\eta_{max} = 1-(T_a/T_b) = 1-300/522$$
$$= 0.425 = 43\% \quad \text{(2 sig. fig.).}$$

The fact that the maximum efficiency is not obtained (23% as against 43%) can be traced to the fact that not all the rejected heat Q_2 is rejected at the lowest temperature (see Carnot's theorem, ex. 17.24).

17.27. Calculate the heat energy rejected per second by a heat engine producing mechanical power of 10 kW with the maximum efficiency possible between a heat source at 200°C and a heat sink at 50°C. [22 J.]

17.28. Calculate the least power required to run a refrigerator which is to abstract heat at a rate of 150 W from a compartment at a constant temperature of −5°C and reject heat at a constant temperature of 40°C. [25 W.]

17.29. A heat pump has a supply of water at a temperature of 10°C and has an output of 150 kW at 25°C. Assuming a coefficient of performance of one-third of the theoretical maximum, calculate he mechanical power required. [23 kW.]

17.30. A heat engine operates in a Carnot cycle as detailed below. Assuming the working substance to be a constant mass of deal gas, having $\gamma = 1.40$, calculate the heat taken in and the work done by the gas in each of the four stages which occur reversibly and in the following order: (1) isothermal expansion at 227°C from pressure of 4.0×10^5 N m^{-2} and volume of 0.10 m^3 o 0.30 m^3; (2) adiabatic expansion to 27°C; (3) isothermal compression at 27°C; (4) adiabatic compression to 227°C. Also calculate the efficiency in two ways: (a) from the heat and work; b) from the temperatures. [(1) $Q = W = 4.4\times10^4$ J; (2) $Q = 0$, $W = 4.0\times10^4$ J; (3) $Q = W = -2.6\times10^4$ J; (4) $Q = 0$, $W = -4.0\times10^4$ J; (a) = (b) = 40%.]

17.31. In an ideal Diesel cycle, air is heated at constant pressure y combustion of the injected fuel, while expansion occurs from volume 5.0×10^{-5} m³ to 10.0×10^{-5} m³. Then further expansion occurs adiabatically to 75×10^{-5} m³. At this point it is assumed that the pressure and temperature drop instantaneously to 1.0×10^{5} N m^{-2} and 300 K, and then the gases in the cylinder are exchanged for an equal volume of fresh air at the same pressure and temperature with no flow of heat or work done. Finally, the fresh air is compressed adiabatically from 75×10^{-5} m³ to 5.0×10^{-5} m³ to complete the cycle. Assuming perfect reversibility, with $\gamma = 1.40$, $c_v = 700$ J kg^{-1} K^{-1}, and $\varrho = 1.29$ kg m^{-3} at s.t.p., calculate (a) the maximum pressure, (b) the maximum temperature, (c) the work done per cycle, and (d) the efficiency. [(a) 44×10^{5} N m^{-2}; (b) 1800 K; (c) 470 J; (d) 61 %.]

INDEX

Absorption of heat 230
Acceleration 1, 46, 180
angular 13, 174–181
centripetal 160–173
gravitational 6–9, 47–50, 166, 203
non-uniform 16–21, 193–211
–time graph of S.H.M. 195
ction and reaction 52, 56
diabatic changes 279–282, 287–290
mplitude of displacement 192
ngle of contact 132–133
rchimedes' principle 115–120
vogadro's law 232–237
vogadro's number 235, 247

alance
of beams 38–45
sensitivity 40
anking of tracks 168–173
arometer 120–122, 130–132, 138
lt drive 14–16, 69–70
nding of beams 41–45
rnoulli's equation 252–258
cycle 83, 172–173
ack body 228–231
ltzmann's constant 246–247
mb calorimeter 159, 224
yle's law 134–138, 141–142, 238, 244
kes 50, 58, 156–157
bbles 127–130

orie 148, 156
orific value of fuel 157–159
pacity, thermal 148–151

Capillary rise 130–133
Carnot's cycle 289
Carnot's theorem 283–285
Celsius scale 97–98, 140, 145–146
Centre of gravity 29–32, 181
Centripetal acceleration and force 160–173
Chain-drive 78–83
Charles' law 139–142
Circle, auxiliary, S.H.M. 194, 200
Collision 7, 53–54, 63, 273
Compressibility 91
Condensation 152–155, 236–238
Conductivity, thermal 212–218
Continuous flow 151, 278
Convection 218, 227
Cooling
correction 224–227
curve 220, 225
Newton's law of 218–225
Coupled objects 56–58, 179
Couples of forces 94, 174
Cycles
of heat engines 267–270, 283–290
of S.H.M. 193

Deceleration 1–5
Density
relative 118–119
vapour 236–242
variation with temperature and pressure 105–107, 143–144, 233–241
Depth gauge 115, 124–126
Dew point 240–241
Diagram, p–V 135, 267–272, 277, 287

Diesel cycle 289–290
Discharge coefficient 254
Displacement–time graph, S.H.M. 192, 195

Efficiency 64, 72, 79, 283–290
Elasticity 84–95, 198–200, 204–208, 211
Emissive power 228
Emissivity 228–231
Energy
  conservation 59, 72
  elastic 94–95, 199
  free surface 129–130
  internal 272–280
  kinetic 59–68, 165, 179–187, 199, 242–247, 253–258, 273–280
  potential 59–61, 163–165, 202, 256–258
  pressure 250
Expansion
  apparent or differential 96, 102, 105–107
  linear 98–103
  real 105–106
  superficial or area 100–103
  volumetric or cubical 103–107, 139
  work by 267–290
Exponential curve 219–224

Fixed points of temperature 97, 145
Flow rate 248–265
Fluid (liquid) 110–133, 248–266
Force
  and acceleration 46, 180
  centripetal 160–173
  combinations of 22–45
  and momentum 51
  parallelogram of 23
  polygon of 23
  resolution of 23, 32, 66
  shearing 41–45, 110
  triangle of 23
Frequency 193, 208, 227
Friction 34–38, 54–58, 63–68, 72–79, 155–157, 168–173
  coefficients 35, 54
  dynamometer 74–75
Fusion, latent heat 152–155

Gas
  constant 233, 247
  equation 141–144, 233
  molecular theory 232–247
  perfect or ideal 134–146, 233–247
  principal specific heats 275–282
Gears 81–83, 178
Gradients 32–34, 54–58, 65–68, 181 184
Gram weight or force 26–45
Gravitation
  law or theory 163, 166
  universal constant 164
Gravity
  acceleration 6–10, 47–49, 166, 20
  centre of 28–32, 181
  force of 48, 163–168
  specific 114, 119
Gun and shell 51–53
Gyration, radius of 175–177, 210

Harmonics 191
Head 112, 257
Heat
  conduction 212–218
  engines 283–290
  latent 152–155
  pump 285–289
  specific 147–152, 275–282
  and temperature 96, 147, 219, 2
  units of 148–150, 155–156
  and work 155–157, 272–290
Hertz (unit) 193
Hooke's law 84–87, 91, 198, 204
Humidity 240–242

Hydraulic jack or press 110, 115
Hydrometer 117–119, 133

Impulse 51–53, 180, 242–243
Inclined plane 32, 54, 65, 181, 184
Isothermal changes 134–138, 271–279

Joule (unit) 60, 148, 155–156
Joule's equivalent 156

Kelvin temperature 140, 144–146
Kilogram force or weight 26, 48
Kilomole 233
Kinetic theory of gas 232–247

Latent heat 152–155, 272–274
Level indicator 115, 124–126
Lifting 69–81
Limit of proportionality 84

Machines 69–83
Manometer 123–127, 136–138
Mass
  relative molecular 232
  and weight 47–49, 166
McLeod gauge 136–138
Mean values in S.H.M. 193, 196
Mechanical advantage 69–74, 77–80
Meniscus 131
Modulus
  bulk 91
  of rigidity 92–94
  Young's 87–91
Molecular attraction 127–130, 152, 236, 242
Moment
  bending 41–45
  of force 30, 94, 171–175
  of inertia
    of bar and block 184–187
    of disc and rod 176–178, 181–184, 187–190
    of sphere 181–183
    of tube and ring 175–178
Momentum 51, 209, 243
  angular 180
  conservation 52
Motion
  angular 13–16, 74–76, 160–190
  parabolic 8–10
  rectilinear 1–12, 16–21
  simple harmonic 191–211

Newton (unit) 23, 26, 46, 48
Newton's law of cooling 218–226
Newton's laws of motion 46, 51, 52, 160

Orbital motion 165–168
Orifice plate 254, 259

Parabolic motion 8
Parallel axes 184–190
Parallelogram
  of forces 23, 25, 32–37
  of velocities 11
Pendulum
  compound 209–211
  simple 21, 201–203
  torsional 206–208, 211
Periodic time 193
Perpendicular axes 187–190
Phase 199
Pitot tube 259
Polygon of forces 23–25
Power 63–68, 150, 180, 217–223, 228–231, 249–252
  in rotation 74–76, 180
Pressure
  atmospheric 112, 120–125
  dynamic 256

Pressure *(cont.)*
fluid 110–146, 232–290
gauge 124, 135–136, 250
kinetic theory of 242–247
mean effective 269–270
partial 234
temperature coefficient of gas 139
vapour 236–242
Progression, geometric 222
Pulleys 14, 69, 77–81

Radian (unit) 13
Radiation 218, 227–231
Range of trajectory 8–10
Refrigerator 285–289
Resultant force 22–25
Reversible changes 283–290
Rigidity 92–95, 110
Rolling 13–16, 181
Rotation 13–16, 74–83, 94–95, 174–208

Satellite 165–168, 231
Shear modulus 92–94
Shell and gun 51–53
Simple harmonic motion 191–211
Sink of heat 284
Sliding *see* Friction
Spring
balance 47, 85–87
helical or spiral 84–87, 95, 204–206
in S.H.M. 198–200, 204–208
Stability of vehicle 168–173
Stefan–Boltzmann law 228–230
Stiffness 85–87, 204–205
Stokes's formula 263
Strain
shear 92–94
tensile 87–90
volumetric 91
Streamline flow 253–265
Stress
shear 93
tensile 87–90
thermal 107–109
volumetric 91
Surface tension 127–133

Temperature
absolute zero 140
ambient 219–223
gradient 212–217
Kelvin (absolute) 140, 144–146, 246–247
kinetic theory 246–247
linear scales 96–98, 144–146
Tension 27, 42, 87, 94, 204
Thermodynamics, first law 273
Thermometers
gas 97, 144–146
mercury-in-glass 96
resistance 97–98
thermoelectric 97
Time
constant 219–223, 266
reaction 5, 51
Torque 74–76, 92–94, 174–176, 180–183, 206–207
Torsion, twisting 92–95, 206–211
Trajectory 8–10
Triangle
of forces 23–27
of velocities 11
Triple point 145–146

U-tube manometer 123–126

Vaporization 152, 236
Velocity 1–21
angular 13–16, 191–196, 208
gradient 260–264
of radiation 227
ratio 72–83
relative 10–12

elocity (*cont.*)
resultant 11–12
root mean square 244–246
terminal 265
–time graph 3, 19, 192
triangle 11–12, 161
enturi tube 252–259
iscometer 261–262
iscosity 259–266
dynamic 261
kinematic 262

Vater equivalent 148, 150–151
Watt (unit) 64, 150
Wavelength 227–231
Weight 26, 47–50, 60–68, 115–120, 166, 201–209
apparent 48–49, 116–117
molecular 232
Weightlessness 166, 168
Weston differential pulley 78–81
Wheel and axle 71–73
Wien's law 228–229
Work 59–75, 94, 155–157, 250–253
and energy 59, 94, 180, 272–290
in expansion 267–290
by flow 250–253
in rotation 94, 180